Standard Letters for Building Contractors

Other books by David Chappell

The JCT Intermediate Building Contracts 2005
Third Edition
David Chappell
978-1-4051-4049-2

The JCT Minor Works Building Contracts 2005
Fourth Edition
David Chappell
978-1-4051-5271-6

The JCT Design and Build Contract 2005
Third Edition
David Chappell
978-1-4051-5924-1

Building Contract Claims
Fourth Edition
David Chappell, Vincent Powell-Smith and John Sims
978-1-4051-1763-0

Parris's Standard Form of Building Contract
Third Edition
David Chappell
978-0-632-02195-6

Building Contract Dictionary
Third Edition
David Chappell, Derek Marshall, Vincent Powell-Smith and Simon Cavender
978-0-632-03964-7

Standard Letters in Architectural Practice
Fourth Edition
David Chappell
978-1-4051-7965-2

Contractual Correspondence for Architects and Project Managers
Fourth Edition
David Chappell
978-1-4051-3514-6

Standard Letters for Building Contractors

Fourth Edition

David Chappell

BA (Hons Arch), MA (Arch), MA (Law), PhD, RIBA
Building Contracts Consultant
David Chappell Consultancy Limited

Blackwell
Publishing

Blackwell Publishing editorial offices:

Blackwell Publishing Ltd, 9600 Garsington Road, Oxford OX4 2DQ, UK
  Tel: +44 (0)1865 776868

Blackwell Publishing Inc., 350 Main Street, Malden, MA 02148-5020, USA
  Tel: +1 781 388 8250

Blackwell Publishing Asia Pty Ltd, 550 Swanston Street, Carlton, Victoria 3053, Australia
  Tel: +61 (0)3 8359 1011

First published 1987 by The Architectural Press
Second edition published 1994 by Blackwell Scientific Publications
Third edition published 2003 by Blackwell Science
Fourth edition published 2008 by Blackwell Publishing Ltd.

ISBN-13: 978-1-4051-7789-4

Library of Congress Cataloging-in-Publication Data

Chappell, David.
    Standard letters for building contractors / David Chappell. – 4th ed.
       p.   cm.
    Includes index.
    ISBN-13: 978-1-4051-7789-4 (hardback : alk. paper)
    1. Construction industry–Law and legislation–Great Britain–Forms.
  2. Construction contracts–Great Britain–Forms.   3. Construction industry–Great
  Britain–Records and correspondence–Forms.   I. Title.

    KD2435.C43 2008
    346.41'07869–dc22

                                                                                    2007025763

A catalogue record for this title is available from the British Library

Set in 10/13 Palatino by Aptara Inc., New Delhi, India

The publisher's policy is to use permanent paper from mills that operate a sustainable forestry policy, and which has been manufactured from pulp processed using acid-free and elementary chlorine-free practices. Furthermore, the publisher ensures that the text paper and cover board used have met acceptable environmental accreditation standards.

For further information on Blackwell Publishing, visit our website:
www.blackwellpublishing.com/construction

Wiley publishes in a variety of print and electronic formats and by print-on-demand. Some material included with standard print versions of this book may not be included in e-books or in print-on-demand. If this book refers to media such as CD or DVD that is not included in the version you purchased, you may download this material at http://booksupport.wiley.com. For more information about Wiley products, visit www.wiley.com

Contents

5 Payment

Letters

Preface to the Fourth Edition

It is good to know that this book is considered useful by the construction industry and especially since the introduction of the CD has made the job of searching for, amending and printing the letters very simple. Building contracts and sub-contracts are constantly being revised and the task of trawling through the many clauses does not become easier. Hopefully, this edition is as clear as the last one.

As well as revising all the letters, omitting some and adding others in the light of case law and legislation, there have been some fundamental changes since the last edition. The most significant change is that the whole suite of JCT contracts was re written and published in 2005. The clause numbers have been completely changed and some things such as Works insurance have been put in schedules at the back of the contract. Terminology has been altered and clauses re-worded. Nominated sub-contractors, performance specified work and the contractor's price statement have been omitted and third party rights have been introduced. The new contracts being dealt with in this book are:

JCT Standard Building Contract (SBC),
JCT Intermediate Building Contract (IC)
    and 'with contractor's design' (ICD),
JCT Minor Works Building Contract (MW)
    and 'with contractor's design' (MWD),
JCT Design and Build Contract (DB).

These replace JCT 98, IFC 98, MW 98 and WCD 98. The new sub-contracts covered by the book are:

JCT Standard Building Sub-Contract Conditions (SBCSub/C)
    and 'with sub-contractor's design' (SBCSub/D/C),
JCT Intermediate Named Sub-Contract Conditions (ICSub/NAM/C),
JCT Intermediate Sub-Contract Conditions (ICSub/C)
    and 'with sub-contractor's design' (ICSub/D/C),
JCT Design and Build Sub-Contract Conditions (DBSub/C).

These replace NSC/C, DSC/C, DOM/2 and NAM/SC.

The GC/Works Sub-Contract has been retained to sit alongside GC/Works/1(1998). It can readily be seen that the number of contracts and sub-contracts (including GC/Works/1) covered by this book has increased from five main contracts and five sub-contracts to seven main contracts and seven sub-contracts. Fortunately, JCT has taken the opportunity to rationalise some clause numbers, which avoids the complications involved in setting out large numbers of alternative clause numbers in the letters. There are now over 300 letters.

In addition to the revised contracts and sub-contracts, the CDM Regulations 2007 were introduced in April 2007, closely followed by JCT Amendment 1 for each contract and sub-contract in the JCT suite which dealt with the new Regulations, changed some of the warranty provisions and amended the termination provisions for IC and ICD. These and the inevitable changes in case law have been taken into account in the letters. Explanatory text is still kept to a minimum in front of each section. As always, it is hoped that the letters, with their titles and text notes, are self-explanatory.

David Chappell
Wakefield
September 2007

Introduction

This book is written for contractors, although it is anticipated that sub-contractors may also find it useful. It consists simply of a set of standard letters for use with the standard forms of building contract and the standard forms of sub-contract. It is not intended to cover unusual situations. A contractor will be involved in writing a good many letters in even the smallest contract if it is carried out under one of the standard forms. Some of the letters will be notices which the contractor must give, others will be letters which it is prudent to send. Composing letters for special circumstances can be a tedious business. Composing them, time after time, for the same standard situation can be a waste of time and effort and, therefore, money.

This book attempts to include all the common situations which a contractor will encounter when involved in a contract. It is, of course, an impossible task and, inevitably, some situations will not be covered. I should be grateful to receive, care of the publishers, any suggestions for standard letters for inclusion in a later edition.

A common criticism of standard, model or typical letters is that the user may copy them for inappropriate situations. Although that could be a danger, it assumes a carelessness on the part of users which, if present, would probably cause them to write equally inappropriate letters of their own composition. To level this criticism is really to misunderstand the purpose of such letters. The user is not relieved of the necessity of understanding contracts or making decisions regarding the content of the letters. I hope they will simply provide a kind of checklist of the letters which should be written, a guide to the matter to be included in each letter and a suggestion for the putting together of words and sentences in particular situations.

Instead of the many directions to, for example, [*insert date*], it would have been possible to put imaginary dates in the letters, quote imaginary companies and cite imaginary situations to illustrate each case. Done like this, the book would have been more fun to write. While not wishing to condemn this approach, which has been used to good effect elsewhere, it seems to me to provide more scope for mistakes in use.

In order to reduce the large number of letters to manageable proportions, the letters have been divided into sections. In the companion volume, *Standard Letters in Architectural Practice*, the RIBA Plan of Work is used as a convenient sequential framework. It is not thought suitable in this instance because the contractor's work does not usually fall into the same divisions.

Instead, a number of key activities have been identified and used to title sections in chronological order: Tendering, Contract Documents, Insurance and

Other Project Planning Matters, Operations on Site, Payment, Extensions of Time, Loss and/or Expense, Termination, Arbitration, Adjudication and Completion, Sub-Contractors. Within each section, the letters have been arranged on the basis that a letter can be found on the first occasion it might be used. It should be possible to locate any particular letter quite easily but, to assist the process, a selective subject index has been incorporated to supplement the complete list of contents. If all else fails, the searchable CD will enable any particular letter to be found.

Unless otherwise stated at the top of each letter, all letters are suitable for use with SBC, IC, ICD, MW, MWD, DB and (GC/Works/1(1998)). Where different contracts require a different letter, a note indicates the fact. Although, in general and for the sake of comparative simplicity, each letter deals with a separate item, it is appreciated that such items are in practice often gathered into one letter.

The section on sub-contractors has been placed at the end of the book, because it is the largest section; it introduces the forms of sub-contract and it tends to mirror the activities in the main contract. Unless otherwise stated at the top of each letter, all letters are suitable for use with SBCSub/C, SBCSub/D/C, ICSub/NAM/C, ICSub/C, ICSub/D/C, DBSub/C and GC/Works/SC. The relationship of the main and sub-contract forms referred to in this book is as follows:

SBC	SBCSub/C
	SBCSub/D/C
IC	ICSub/NAM/C*
	ICSub/C
ICD	ICSub/NAM/C*
	ICSub/D/C
DB	DBSub/C
MW	—
MWD	—
GC/Works/1	GC/Works/SC

* This is the same contract.

At the time of writing, there is no standard form of sub-contract for use with MW or MWD.

The following points should be borne in mind when using this book:

- When referencing SBC, the With Quantities version has been used. Although most of the letters can be used for the other two versions, care must be taken to check before use, because there are differences in some instances.
- Every letter should carry a heading giving the project title. Headings have been omitted here for simplicity's sake.
- It has been assumed throughout that the contractor and sub-contractors are corporate bodies and that the contractor acts as 'principal contractor' under the CDM Regulations 2007. It is common for the contractor also to act as CDM co-ordinator under DB.

- Variants to suit different contracts are usually given in the same letter, but if it is more convenient, less confusing or an important letter is concerned, separate alternative letters are given.
- The terms 'architect', 'client' or 'employer' have been used throughout for consistency, but it should be noted that when using GC/Works/1(1998) or GC/works/SC, the architect is termed the 'PM' and when using DB, there is no architect in a traditional role and letters normally sent to the architect should be addressed to the 'employer's agent', if appointed, otherwise to the 'employer'.
- It has been assumed that the Supplemental Provisions are used with DB.
- Standard documents available elsewhere, such as standard forms of tender and certificates of various kinds, are not included.
- Letters which a contractor might write with regard to VAT, the Finance Act and fluctuation matters have not been included, because they are liable to sudden change and/or they are not suitable for treatment by standard letter in most instances.

A word of warning: standard letters can be very useful to a busy contractor, but they can also be very dangerous if they are used without thought. Always consider carefully whether a standard letter is really appropriate in the circumstances. If in doubt, seek advice.

Companion website

This book is accompanied by a companion website:

http://www.wiley.com/legacy/wileychi/chappell1

The website includes:

- The preface
- The introduction
- All the standard letters found in this book
- The index

In order to gain access to the downloadable material, please go to the web address above and fill in your details on the web form. Proof of purchase will also be required to gain access to the downloadable letters – please email your receipt separately to: standardlettersforbuildingcontractors@wiley.com.

Please visit the website for further information.

---

1   Tendering

There are various standard published forms which are commonly used in connection with tendering procedures. The *Code of Procedure for Single Stage Selective Tendering 1996*, the *Code of Procedure for Two Stage Selective Tendering 1996* and the *Code of Procedure for Selective Tendering for Design and Build 1996* used to be used, but it is now usual for the *Code of Practice for the Selection of Main Contractors 1997*, prepared by the Construction Industry Board, to be employed.

The tender stage is normally the first contact between contractor and architect. The stage can be very long and frustrating, particularly if programmed dates for delivery of tender documents are missed, as happens all too frequently. Where a contractor is being asked to tender for design and build on the basis of JCT contract DB the tender period is likely to be much longer than under the traditional system of procurement. This is because the contractor must have additional time to produce design proposals before getting down to preparing a price. In order to avoid a serious waste of time and money, it is common to invite such tenders on a two stage basis.

Tendering is the contractor's way of obtaining work and a persistent attitude to the initial hurdle of getting on the tender list should be adopted. Many local authorities maintain lists of contractors. Private employers are often advised by the architect and the quantity surveyor with regard to the firms to be included on the tender list. It is, however, a decision for the employer to make. The first letter is designed for situations where you become aware that a project is at design stage. Although it is addressed to the architect, there may be merit, in some cases, in addressing it to the employer if you consider that the architect is unlikely to act on it. If such a letter produces no reply, it is always worthwhile to follow it up.

Letter 1
To architect, requesting inclusion in list of tenderers

Dear

We were interested to note from [*state source*] that tenders are to be invited for the above project in the near future.

This is a type of work in which we are very experienced and we should welcome an invitation to tender. The following information may be of assistance to you if you are not already aware of our capabilities:

[*List the following information:*

1. *Names and addresses of all directors.*
2. *Address of registered office.*
3. *Website.*
4. *Share capital of firm.*
5. *Annual turnover during the last three years.*
6. *Number and positions of all office-based staff.*
7. *Number of site operatives permanently employed in each trade.*
8. *Number of trained supervisory staff.*
9. *Number and value of current contracts on site.*
10. *Address, date of completion and value of three recently completed projects of similar character to that for which tenders are to be invited.*
11. *Names and addresses of clients, architects or quantity surveyors connected with the projects noted in 10 above and to whom reference may be made.*]

We look forward to hearing from you in due course.

Yours faithfully

Letter 2
To architect, if no response to request for inclusion on list of tenderers

Dear

We refer to our letter of the [*insert date*], requesting inclusion on the list of tenderers for the above project.

Since we have not heard from you, we take this opportunity to re-affirm our interest in the project and assure you of our experience in work of this nature.

We should be delighted to meet you to expand upon the details given in our earlier letter. We have the facilities to make a presentation showing recent projects we have carried out which may be of interest to your client.

Our managing director [*or insert appropriate designation*], M.. [*insert name*], will telephone you on [*insert day*].

Yours faithfully

Letter 3
To architect, agreeing to tender

Dear

Thank you for your letter of the [*insert date*] from which we note that you intend to invite tenders for the above project.

We should be pleased to be included on the tender list. No doubt you will be sending further details in due course.

Yours faithfully

Letter 4
To architect, if contractor unwilling to tender

Dear

Thank you for your letter of the [*insert date*] from which we note that you intend to invite tenders for the above project.

With regret, we must ask to be excused from tendering on this occasion due to our very heavy workload. We do hope, however, that you will give us the opportunity to tender for other projects on other occasions in the future.

Yours faithfully

Letter 5
To architect, if contractor asked to provide information prior to inclusion on tender list

Dear

Thank you for your letter of the [*insert date*] setting out brief details of the above project and requesting particulars of this company.

On the information provided, it appears that the project would be directly related to our skills and experience and we should be delighted to be included on the tender list. The information you require is as follows:

[*List answers using same numeration as in architect's letter.*]

Yours faithfully

Letter 6
To architect, if the contractor is informed that the tender date is delayed and is still willing to submit tender

Dear

Thank you for the letter of the [*insert date*] informing us that the date for despatch of tender documents has been revised to [*insert date*]. We confirm that we are still willing to submit a tender for this project.

Yours faithfully

Letter 7
To architect, if the contractor is informed that the tender date is delayed and is unwilling to tender

Dear

Thank you for your letter of the [*insert date*] informing us that the date of despatch of tender documents has been revised to [*insert date*]. We regret that it will be impossible for us to rearrange our very heavy workload so as to be able to submit a tender in accordance with the new timetable.

We do hope, however, that you will give us the opportunity to tender for other projects on other occasions in the future.

Yours faithfully

Letter 8
To architect, acknowledging receipt of tender documents

Dear

Thank you for your formal invitation to tender for the above project with which you enclosed [*list documents enclosed*].

We confirm that we will submit our tender by the [*insert tender date*].

[*If appropriate, add:*]

We wish to inspect the detailed drawings and visit site. Our M.. [*insert name*] will telephone you to make the necessary appointment within the next few days.

Yours faithfully

Letter 9
To architect, regarding questions during the tender period

Dear

We have carefully examined the tender documents enclosed with your letter of the [*insert date*]. We have examined the detailed drawings at your office and we have visited site. There are certain items which require clarification as follows:

[*List items requiring clarification.*]

Items marked with a red X are urgent and, if we are to meet the date for submission of tenders, we need clarification of these points by [*insert date*].

Yours faithfully

Letter 10
To architect, requesting extension of tender period

Dear

We are preparing our tender for the above project with the greatest possible speed. Prices for a number of the sub-contract items, however, will not be in our hands until after the date for submission of tenders. Clearly, we will be unable to submit a tender unless the tendering period is extended. We therefore request an extension of the period by [*insert period of extension, which should be as short as possible*]. We are proceeding on the assumption that you will be able to grant our request, but if you feel unable so to do, please let us know immediately so that we can stop what will become abortive work.

Yours faithfully

Letter 11
To architect, withdrawing qualification to tender

Dear

In response to your letter of the [*insert date*], we confirm that we withdraw the qualification to our tender dated [*insert date*] without amendment to the tender sum of [*insert amount*].

The qualification to which we refer above is:

[*Set out the qualification using the precise wording used in the tender.*]

Yours faithfully

Letter 12
To architect, if confirming offer where the overall price is dominant

Dear

Thank you for your letter of the [*insert date*] with which you enclosed a list of errors detected in our pricing of the bills for the above project.

We have carefully examined the list and we note that, in accordance with the CIB Code of Practice for the Selection of Main Contractors, the overall price is to be dominant in this instance. Therefore, please take this as notice that we confirm our offer of [*insert amount*] as stated in our tender dated [*insert date*].

We note what you state regarding endorsement and we agree to its terms.

Yours faithfully

Letter 13
To architect, if withdrawing offer where the overall price is dominant

Dear

Thank you for your letter of the [*insert date*] with which you enclosed a list of errors detected in our pricing of the bills for the above project.

We have carefully examined the list and we note that, in accordance with the CIB Code of Practice for the Selection of Main Contractors, the overall price is to be dominant in this instance. In view of the nature of the errors, we regret that we must withdraw our offer.

Yours faithfully

Letter 14
To architect, if amending offer where the pricing document is dominant

Dear

Thank you for your letter of the [*insert date*] with which you enclosed a list of errors detected in our pricing of the bills for the above project.

We have carefully examined the list and we note that, in accordance with the CIB Code of Practice for the Selection of Main Contractors, the pricing document is dominant in this instance. In view of the nature of the errors, we have amended our offer. Our amended tender price is [*insert amount*] and we enclose details of the relevant calculations.

Yours faithfully

Letter 15
To architect, if tender accepted (a)
This letter is not suitable for use with DB, ICD or MWD.

Dear

Thank you for your letter of the [*insert date*] accepting our tender of the [*insert date*] in the sum of [*insert amount*] for the above Works in accordance with the drawings numbered [*insert numbers*] and the bills of quantities [*or specification/work schedules*].

We understand that a contract now exists between the employer and ourselves and we look forward to receiving the contract documents for signing/execution as a deed [*delete as appropriate*] in due course.

Yours faithfully

Letter 16
To architect, if tender accepted (b)
This letter is only suitable for use with DB.

Dear

Thank you for your letter of the [*insert date*] accepting our tender of the [*insert date*] in the sum of [*insert amount*] for the completion of the design and the construction of the above project in accordance with the Employer's Requirements, the Contractor's Proposals and the Contract Sum Analysis.

We understand that a contract now exists between the employer and ourselves and we look forward to receiving the contract documents for signing/execution as a deed [*delete as appropriate*] in due course.

Yours faithfully

Letter 17
To architect, if tender accepted (c)
This letter is only suitable for use with ICD or MWD.

Dear

Thank you for your letter of the [*insert date*] accepting our tender of the [*insert date*] in the sum of [*insert amount*] for the construction of the above project and the design and construction of the Contractor's Designed Portion in accordance with the drawings numbered [*insert numbers*] and the bill of quantities [*or specification/work schedules*] and the relevant Employer's Requirements, Contractor's Proposals and CDP Analysis.[1]

We understand that a contract now exists between the employer and ourselves and we look forward to receiving the contract documents for signing/execution as a deed [*delete as appropriate*] in due course.

Yours faithfully

[[1] *When using MWD, delete 'Contractor's Proposals and CDP Analysis'.*]

Letter 18
To architect, if purporting to accept tender

Dear

Thank you for your letter of the [*insert date*], which purports to be an acceptance of our tender of the [*insert date*] for the Works.

It is clear that your letter is not an acceptance of our tender, indeed it attempts to insert the following changes: [*list the changes*].

Therefore, your letter is, strictly, a counter-offer and no contract exists unless we accept it. In the circumstances, we have no intention of accepting it. We are advised that your counter-offer had the effect of rejecting our original tender. Therefore, we confirm that we will re-open our tender for a further [*insert period*] to allow you to unequivocally accept it.

Yours faithfully

Letter 19
To architect, if another tender accepted

Dear

Thank you for your letter of the [*insert date*] from which we note that our tender was not successful in this instance.

We await details of the full list of tender prices with interest and assure you of our willingness to receive your future enquiries.

Yours faithfully

2 Contract Documents

If a contractor is successful at tender stage, the architect or the employer will write, informing you of the fact. Such a letter may or may not be an acceptance which will form a binding contract. An unequivocal letter of acceptance is what you are looking for, but it is a rare commodity. If the letter is hedged around with provisos or made subject to agreeing further terms, it will not usually form a contract.

The letters which follow cover the basic situations which might arise in relation to the contract documents which will be prepared for signature or for completion as a deed by the architect or, in the case of a local authority, by the legal department. It is no longer necessary to seal a document in order to make it a deed, although you may continue to seal if you so wish. In view of the differing limitation periods (6 and 12 years) which will apply, depending on whether the document is under hand or a deed, it is wise to get advice if you are in any doubt about the procedure or the consequences. In any event, it is crucial to examine the documents with great care.

The printed contract itself should be checked against the information given in the tender documents and the discrepancies noted. The contract drawings must be identical with the drawings on which you tendered. Sometimes, in a traditional procurement situation, such drawings are revised by the architect between invitation to tender and the completion of the contract documents. If you find any inconsistencies at all, refuse to sign the documents until they are corrected, no matter that the architect assures you that they do not affect the contract. Particular care must be exercised when you enter into a contract on the basis of DB. The Employer's Requirements and the Contractor's Proposals must be consistent, because the contract does not contain any obvious mechanism for correcting discrepancies between the two documents. There will still be a danger that the original drawings forming the basis of your tender will be amended before the formal documents are put together, but the person responsible is more likely to be you than the employer's agent. It is not sufficient merely to list the inconsistencies between the Employer's Requirements and the Contractor's Proposals. They must be amended so as to be consistent. This is very important because, if not amended, the Employer's Requirements will usually take precedence.

Some of the letters deal with the circumstances if work commences on site before the documents are complete. This situation should not be allowed to occur. In practice, it often happens. There will be little difficulty provided that a binding contract is in existence and that it precisely reflects the terms of the formal documents. This it will often do by incorporation, i.e. specifically referring to them. Provided such terms are incorporated, there appears to be no reason why the architect cannot, for example, issue instructions and certificates.

Rather than having to write the sort of letter suggested if certification is due, you should insist on the contract documents being prepared before taking possession of the site. There is seldom any good reason why they should not be ready for execution immediately after your tender is accepted. Remember, however, that the architect's failure to prepare contract documents is not a sufficient reason for you to refuse to take possession, provided a proper binding contract incorporating all the terms of the formal documents has been entered into.

Letter 20
To architect, returning contract documents
Special delivery

Dear

Thank you for your letter of the [*insert date*] with which you enclosed the contract documents. We have pleasure in returning them herewith, duly signed/executed as a deed [*delete as appropriate*] as requested.

We look forward to receiving a certified copy of the contract documents within the next few days.

Yours faithfully

Letter 21
To architect, if mistakes in contract documents and previous acceptance of tender

Dear

We are in receipt of your letter of the [*insert date*] with which you enclosed the contract documents for us to sign/execute as a deed [*delete as appropriate*].

There is an error on [*describe nature of error and page number of document*]. This is not consistent with the tender documents on which our tender is based which, together with your acceptance of the [*insert date*], forms a binding contract between the employer and ourselves.

We therefore return the documents herewith and we look forward to receiving the corrected documents as soon as possible.

Yours faithfully

Letter 22

To architect, if mistakes in contract documents and no previous acceptance of tender

Dear

We are in receipt of your letter of the [*insert date*] with which you enclosed the contract documents for us to sign/execute as a deed [*delete as appropriate*].

There is an error on [*describe nature of error and page number of document*]. This is not consistent with the tender documents on which our tender is based and we are not prepared to enter into a contract on the basis of the contract documents in their present form.

We therefore return the documents herewith and we look forward to receiving the corrected documents as soon as possible.

Yours faithfully

Letter 23

To architect, if contractor asked to commence before contract documents signed, but tender accepted

Dear

Thank you for your letter of the [*insert date*] from which we note that the employer requests us to commence work on site pending completion of the contract documents.

It is our understanding of the situation that we are already in a binding contract with the employer on the basis of our tender of the [*insert date*] and an acceptance of the [*insert date*] on terms incorporated by such tender and letter of acceptance.

If the employer will send us written confirmation indicating agreement with our understanding of the situation as expressed in this letter, we will be happy to commence as requested.

Yours faithfully

Letter 24
To architect, if contractor asked to commence before contract documents signed and tender not accepted

Dear

Thank you for your letter of the [*insert date*] from which we note that the employer requests us to commence work on site pending completion of the contract documents.

We note that the employer has not yet accepted our tender of the [*insert date*] and, therefore, there is currently no contract between us, no mechanism for payment, instructions etc. We are sure that you will understand our reluctance to commence work under these circumstances.

Obviously, if the employer will send us written and unequivocal acceptance of our tender or executes the contract documents, we will be happy to commence as requested.

Yours faithfully

Letter 25
To architect, if contract not signed and certification due
This letter is not suitable for use with DB.
Special delivery

Dear

We tendered for the above work on the [*insert date*] in the sum of [*insert amount*] and the employer accepted our tender unequivocally by letter dated the [*insert date*]. A binding contract exists between the employer and ourselves in terms incorporating, among other things, the provisions of [*insert the full title of the form of contract*] which provides for certification of monies due to us.

Although we have difficulty in understanding why you have not yet prepared the formal contract documents for signature/completion as a deed [*delete as appropriate*], the completion of such formal contract documents will simply reflect the respective rights and obligations of the parties as already agreed and about which there is no doubt.

We, therefore, require you to issue your certificate in accordance with clause 4.9 [*substitute '4.8' when using IC or ICD, or '4.3' when using MW or MWD, or '50' when using GC/Works/1 (1998)*] of the conditions of contract

If such certificate is not in our hands by [*insert date*] we will take immediate legal action against the employer for the breach.

Yours faithfully

3 Insurance and Other Project Planning Matters

The letters which follow cover the insurance clauses in the standard forms. With the exception of GC/Works/1, they are all very similar. MW and MWD are somewhat shorter than the other JCT provisions, as one might expect. Minor differences between the forms should be noted. If you already maintain insurances of a general nature, they may be acceptable provided that they cover the risks and the sums adequately. Even where contractor's all risks insurance is held on an annual basis, you must make sure, at the very least, to report each new project to the insurers. Insurance is a specialist subject which is best dealt with by your broker.

If it is for the employer to take out insurance; for example in the case of existing buildings and works of extension, you must be sure to inspect such insurances before commencing on site. Remember that insurers tend to interpret 'existing structures' quite strictly and this insurance must be used even if there is but one small piece of wall to be retained from the existing property. In general, local authorities are not required to insure, presumably on the basis that it is more cost effective to meet any risks from their own resources.

Since the previous major revision of the JCT contracts in 1980, the insurance provisions have been frequently amended, ostensibly to take account of decisions from the courts.

Third party rights and collateral warranties deserve (and have been given) a book to themselves. Letters suggesting some responses to requests to enter into warranties or performance bonds have been included. Only if the warranty's terms were among the documents on which you tendered are you bound to complete it. Letters will also be required for the appointment of the site agent, manager or foreman.

Letter 26
To architect, seeking approval to the names of insurers for employer's liability
This letter is not suitable for use with MW, MWD OR GC/Works/1 (1998).

Dear

In accordance with the provisions of clause 6.5.2 and to avoid any delays to the progress of the Works, we request the employer's approval by [*insert date*] to the following insurers to provide clause 6.5.1 insurance: [*insert name and address of the insurers*].

[*Or:*]

In accordance with the provisions of Schedule 3 [*substitute '1' when using IC or ICD*], paragraph A.1 and to avoid any delays to the progress of the Works, we request the employer's approval by [*insert date*] to the following insurers to provide paragraph A.1 insurance: [*insert name and address of the insurers*].

[*Or:*]

In accordance with the provisions of clause 6.5.2 and Schedule 3 [*substitute '1' when using IC or ICD*], paragraph A.1 and to avoid any delays to the progress of the Works, we request the employer's approval by [*insert date*] to the following insurers to provide clause 6.5.1 and paragraph A.1 insurance respectively: [*insert name and address of the insurers*].

Yours faithfully

Letter 27
To architect, regarding contractor's insurance (a)
This letter is not suitable for use with MW, MWD or
GC/WORKS/1 (1998).
Special delivery

Dear

Thank you for your letter of the [*insert date*]. In response to your requirement,
we have pleasure in enclosing documentary evidence in accordance with
clause 6.4.2 of the contract for inspection by the employer. Please confirm that
it is to the employer's satisfaction.

[*If appropriate, add either:*]

In accordance with Schedule 3, paragraph A.2 [*substitute 'Schedule 1' when
using IC or ICD*], we hereby submit for deposit with the employer the joint
names policy referred to in paragraph A.1.

[*Or:*]

In accordance with Schedule 3, paragraph A.3 [*substitute 'Schedule 1' when using
IC or ICD*] we enclose documentary evidence that we maintain appropriate All
Risks cover independently of our obligations under this contract. Please note,
for insertion in the appendix, that the annual renewal date is [*insert date*].

Yours faithfully

Letter 28
To architect, regarding contractor's insurance (b)
This letter is only suitable for use with MW or MWD.
Special delivery

Dear

Thank you for your letter of the [*insert date*]. In response to your requirement, we have pleasure in enclosing documentary evidence that the insurances required under clause 5.3 have been taken out and are in force at all material times.

[*If appropriate, add either:*]

We also enclose documentary evidence that the insurance required under clause 5.4A of the contract has been taken out and is in force at all material times.

[*Or:*]

We enclose documentary evidence that we maintain appropriate All Risks cover independently of our obligations under this contract.

[*Then add:*]

Please confirm that these insurances are all to the employer's satisfaction.

Yours faithfully

Letter 29

To employer, regarding contractor's insurance (c)
This letter is only suitable for use with GC/Works/1 (1998).
Special delivery

Dear

Thank you for your letter of the [*insert date*]. In response to your request, we have pleasure in enclosing copies of the following insurance policies in accordance with clause 8 of the conditions of contract.

[*List policies together with numbers and dates.*]

Yours faithfully

Letter 30

To employer, within 21 days of acceptance of tender or renewal of insurance
This letter is only suitable for use with GC/Works/1 (1998).
Special delivery

Dear

In accordance with clause 8(4) of the conditions of contract, we have pleasure in enclosing a certificate from our insurer/broker [*delete as appropriate*] attesting that the appropriate insurance policies have been effected.

Yours faithfully

Letter 31
To architect, after approval of insurers for employer's liability
This letter is not suitable for use with MW, MWD OR
GC/WORK/1 (1998).
Special delivery

Dear

We enclose insurance policy number [*insert number*] and premium receipt dated [*insert date*] in respect of insurance taken out with approved insurers, [*insert name*], for deposit with the employer in accordance with clause 6.5.2 in joint names.

Yours faithfully

Letter 32
To architect, regarding professional indemnity insurance
This letter is not suitable for use with IC, MW or MWD.

Dear

Further to your letter dated [*insert date*] requesting documentary evidence that the insurance required under clause 6.11 [*substitute '6.15' when using ICD or '8A' when using GC/Works/1 (1998)*] is being maintained, we have pleasure in enclosing the following documents [*list*].

[*If any original documents have been included, send the letter by special delivery and add:*]

Please note that we have included the originals of documents [*identify*] and we should be grateful if you would return them, using special delivery, by [*insert date*].

Yours faithfully

Letter 33
To architect, if professional indemnity insurance is no longer available at commercially reasonable rates
This letter is not suitable for use with IC, MW or MWD.

Dear

In attempting to renew our professional indemnity insurance, we have to inform you in accordance with clause 6.12 [*substitute '6.16' when using ICD or '8A(2)' when using GC/Works/1 (1998)*] that the insurance referred to in clause 6.11 [*substitute '6.15' when using ICD or '8A' when using GC/Works/1 (1998)*] is no longer available at commercially reasonable rates.

We should be glad to discuss the matter with you if you will let us know a suitable time and date, bearing in mind that the insurance renewal date is [*insert date*].

Yours faithfully

Letter 34
To architect, if Joint Fire Code remedial measures are a variation
This letter is only suitable for use with SBC, IC and ICD.

Dear

We hereby give notice as required by clause 6.15.1.2 [*substitute '6.13.1.2' when using IC or ICD*] that the Remedial Measures as required by the insurers under the provisions of clause 6.15.1.1 [*substitute '6.13.1.1' when using IC or ICD*] require a variation.

Please issue such instructions as are necessary to enable compliance.

Please take this letter as a notice of delay under clause 2.27.1 [*substitute '2.19.1' when using IC or ICD*] involving a Relevant Event under clause 2.29.1 [*substitute '2.20.1' when using IC or ICD*] and an application for loss and/or expense under clause 4.23 [*substitute '4.17' when using IC or ICD*] involving a Relevant Matter under clause 4.24.1 [*substitute '4.18.1' when using IC or ICD*].

Yours faithfully

Letter 35
To architect, if Joint Fire Code remedial measures are a variation, but require emergency action
This letter is only suitable for use with SBC, IC and ICD.

Dear

We hereby give notice as required by clause 6.15.1.2 [*substitute '6.13.1.2' when using IC or ICD*] that we have been obliged to commence the carrying out of work constituting emergency compliance with Remedial Measures as required by the insurers under the provisions of clause 6.15.1.1 [*substitute '6.13.1.1' when using IC or ICD*]. The steps we are taking to supply limited materials and execute limited work are as follows:

[*Insert details.*]

We confirm that the work carried out and the materials supplied rank as a variation. Please provide further instructions necessary.

Please take this letter as a notice of delay under clause 2.27.1 [*substitute '2.19.1' when using IC or ICD*] involving a Relevant Event under clause 2.29.1 [*substitute '2.20.1' when using IC or ICD*] and an application for loss and/or expense under clause 4.23 [*substitute '4.17' when using IC or ICD*] involving a Relevant Matter under clause 4.24.1 [*substitute '4.18.1' when using IC or ICD*].

Yours faithfully

Letter 36

To employer, regarding employer's insurance (a)
This letter is only suitable for use with SBC and DB.

Dear

We should be pleased to receive documentary evidence and receipts showing that the joint names insurance policy has been taken out and is being maintained in accordance with Schedule 3, paragraphs B.2.1.1/C.3.1.1 [*delete as appropriate*] of the contract. The receipt should indicate that such insurance is currently effective and it should be in our hands by [*insert date*].

Yours faithfully

Letter 37
To employer, regarding employer's insurance (b)
This letter is only suitable for use with IC and ICD.

Dear

We should be pleased to receive documentary evidence and receipts showing that the joint names insurance policy has been taken out and is being maintained in accordance with Schedule 1, paragraphs B.2.1.1/C.3.1.1 [*delete as appropriate*] of the contract. The receipt should indicate that such insurance is currently effective and it should be in our hands by [*insert date*].

Yours faithfully

Letter 38
To employer, regarding employer's insurance (c)
This letter is only suitable for use with MW or MWD.

Dear

We should be pleased to receive documentary evidence showing that the joint names insurance policy has been taken out and is being maintained in accordance with clause 5.4B/5.4C [*delete as appropriate*] of the contract. The evidence should indicate that such insurance is currently effective and it should be in our hands by [*insert date*].

Yours faithfully

Letter 39
To employer, who fails to maintain insurance cover (a)
This letter is not suitable for use with MW or MWD or GC/Works/1 (1998).

Dear

We refer to your telephone conversation with our M.. [*insert name*] today and we confirm that you are unable to produce a receipt, as required by Schedule 3 [*substitute 'Schedule 1' when using IC or ICD*], paragraph B.2.1.1/C.3.1.1 [*delete as appropriate*], showing that insurance is currently effective.

In view of the importance of the insurance and without prejudice to your liabilities, we are arranging to exercise our rights under the above mentioned clause immediately. On production of a receipt for any premium paid, we will be entitled under paragraph B.2.1.2/C.3.1.3 [*delete as appropriate*] to have its amount added to the contract sum.

[*If a failure to insure under paragraph C.1, add:*]

We draw your attention to the fact that we intend to exercise our right of entry and inspection for the purposes of survey and inventory on [*insert date*] at approximately [*insert time*].

Yours faithfully

Copy: Architect

Letter 40
To employer, who fails to maintain insurance cover (b)
This letter is only suitable for use with MW or MWD.

Dear

We refer to your telephone conversation with our M.. [*insert name*] today and we confirm that you are unable to produce evidence, as required by clause 5.5, that the insurance referred to in clause 5.4B/5.4C [*delete as appropriate*] has been taken out and is in force at all material times.

You are in breach of contract for which we intend to claim damages. We are advised that such damages include our costs in taking out appropriate insurance.

Yours faithfully

Letter 41
To architect and employer, if any damage occurs due to an insured risk (a)
This letter is only suitable for use with SBC or DB.

Dear

In accordance with Schedule 3, paragraph A.4.1/B.3.1/C.4.1 [*delete as appropriate*] of the conditions of contract, we hereby give notice that loss/damage [*delete as appropriate*] has been caused to the work executed/site materials [*delete as appropriate*] by one of the risks covered by the joint names policy.

[*Describe the loss or damage, stating extent, nature and location.*]

Yours faithfully

Letter 42

To architect and employer, if any damage occurs due to an insured risk (b)
This letter is only suitable for use with IC or ICD.

Dear

In accordance with Schedule 1, paragraph A.4.1/B.3.1/C.4.1 [*delete as appropriate*] of the conditions of contract, we hereby give notice that loss/damage [*delete as appropriate*] has been caused to the work executed/site materials [*delete as appropriate*] by one of the risks covered by the joint names policy.

[*Describe the loss or damage, stating extent, nature and location.*]

Yours faithfully

Letter 43
To architect, on receipt of letter of intent

Dear

We are in receipt of a letter of intent dated [*insert date*].

[*If not prepared to proceed, add:*]

The tender documents gave no indication that you might ask us to proceed on the basis of a tender of intent and we are not prepared to do so. It is far too risky for all parties. We will proceed on the basis of an unequivocal acceptance of our tender or on the execution of the formal contract documents.

[*If prepared to proceed, add:*]

Working on the basis of a letter of intent is risky for both parties. We are prepared to commence work on the basis of your letter, but note that we expect formal contract documents to be prepared ready for execution by both parties within the next [*insert period*]. Failure to do so will cause us to reassess the situation at that stage and probably to cease all further work.

Yours faithfully

Letter 44
To architect, if contractor asked to sign a warranty

Dear

Thank you for your letter of the [*insert date*] with which you enclosed a form of warranty for signature.

[*Add either:*]

We note that the form is identical to that attached to the contract documents and we have pleasure in returning it duly completed as requested.

[*Or:*]

Although the contract documents call on us to complete a form of warranty, no particular form was specified and we regret that your suggested form is not acceptable.

[*Or:*]

We note that there was no requirement in the contract documents for us to complete a warranty. However, we enclose a suggested warranty which we would be prepared to complete if you will complete a warranty to us as the form attached.

Yours faithfully

Letter 45
To architect, on receipt of defective third party rights notice
This letter is only suitable for use with SBC or DB.

Dear

Thank you for your letter dated [*insert date*] which purports to be a notice under clause 7A.1 [*substitute '7B.1' if the notice refers to a funder*].

[*Add either:*]

Please note that the purchaser/tenant/funder [*delete as appropriate*] named in your notice cannot acquire third party rights under this contract because such purchaser/tenant/funder was not identified by name, class or description in part 2 of the contract particulars.

[*Or:*]

Please note that your notice is defective in that [*set out why it is defective*]. Consequently, the purchaser/tenant/funder [*delete as appropriate*] named in your notice cannot acquire third party rights under this contract.

Yours faithfully

Letter 46
To architect, on receipt of request for warranty
This letter is not suitable for use with MW, MWD or GC/Works/1 (1998).

Dear

Thank you for your letter dated [*insert date*] issued as a notice under clause 7C/7D/7E [*substitute 7.5/7.6/7.7 when using IC or ICD and delete as appropriate*].

[*Add either:*]

However, we should point out that we are not required to provide such a warranty/sub-contractor warranty [*delete as appropriate*] because the relevant details were not inserted in part 2 of the contract particulars.

[*Or:*]

We are putting the necessary measures in place to provide the warranty as requested.

Yours faithfully

Letter 47
To architect, if contractor asked to supply a performance bond

Dear

Thank you for your letter of the [*insert date*] requesting us to provide a performance bond.

[*Add either:*]

We note that a form of bond was included in the contract documents and we are arranging for a bond to be executed accordingly.

[*Or:*]

Although the contract documents call on us to provide a performance bond, no particular form was specified. The form you include with your letter is not acceptable and we are arranging to have the bond executed in a form which we find acceptable.

[*Or:*]

The contract documents do not require us to provide a performance bond and we decline to do so.

Yours faithfully

Letter 48

To employer, if asked to execute a novation agreement
This letter is only suitable for use with DB.

Dear

Thank you for your letter dated [*insert date*] with which you enclosed a novation agreement which you are asking us to execute. By this agreement, we would effectively take over the contract you have with the architect [*or substitute the relevant consultant*] just as though he/she had been in contract with us from the beginning.

[*Add either:*]

This is the same document that we undertook to execute when we tendered for this project; therefore, we have had the relevant signatures appended and we return it herewith. Although we have kept a copy, we should be glad to receive the final version properly executed by all parties.

[*continued*]

Letter 48 continued

[*Or:*]

Although we undertook to execute a novation agreement, we did not bind ourselves to execute any particular agreement which you might chose to present to us. We have not had the opportunity to see this agreement before and, therefore, we have requested legal advice on the content. We will let you know the result in due course. In the meantime any delay in appointing the architect [*or substitute the name of the consultant*] will delay the start of the Works on site and we shall expect an appropriate extension of the contract period and reimbursement of our direct loss and/or expense, alternatively, damages at common law.

[*Or:*]

We have never undertaken to execute a novation agreement and we decline to do so now. Our intention is to enter into a completely separate agreement with the architect [*or substitute the name of the consultant*].

Yours faithfully

Letter 49
To architect, enclosing the construction phase plan

Dear

[*If using SBC, IC, ICD, MW, MWD or DB begin:*]

The CDM planning period ends on the [*insert date*].

[*Then continue, or otherwise, begin:*]

We have completed the construction phase plan so far as we are able on the basis of the information currently to hand. The purpose of the plan is to show the way in which the construction phase is to be managed and the important health and safety issues in this project.

We take the opportunity to enclose a full copy of the plan for your comments and use. Please note that it is not to be treated as a mere paper exercise. Rather, it is an important tool in the construction process and something which is a mandatory requirement under the Construction (Design and Management) Regulations 2007. You will see that the plan is divided into the following main parts:

- description of the project,
- management of the work,
- arrangements for controlling significant site risks,
- health and safety file.

[*continued*]

Letter 49 continued

Copies of this letter and the complete plan are being sent to the client and other consultants and relevant parts of the plan to other parties as necessary.

Yours faithfully

Copies: Employer
CDM Co-ordinator [unless the contractor takes this role]
Other consultants

Letter 50
To architect, regarding person-in-charge or agent (a)

Dear

This is to inform you that the person-in-charge [*substitute 'agent' when using GC/Works/1 (1998)*] of this project on site will be [*insert name*]. He/She [*delete as appropriate*] is competent and experienced in this type of work.

Please report your presence on site to the person-in-charge/agent [*delete as appropriate*] immediately on arrival for health, safety and security reasons.

Yours faithfully

Letter 51
To architect, regarding the appointment of a site manager (b)
This letter is only suitable for use with DB.

Dear

We propose to appoint [*insert name*] as site manager for this project. He/She [*delete as appropriate*] is competent and experienced in this type of work and we should be pleased to receive your written consent to this appointment as required by Schedule 2, supplemental provision paragraph 1.2 of the contract.

Please report your presence on site to the site manager immediately on arrival for health, safety and security reasons.

Yours faithfully

Letter 52
To architect, regarding consent to removal or replacement of site manager
This letter is only suitable for use with DB.

Dear

[*If removal:*]

We refer to our telephone conversation of the [*insert date*] and confirm that we intend to remove [*insert name*] as site manager for the reasons we discussed. We understand that you accept those reasons and we should be pleased to receive your written consent, as required by Schedule 2, supplemental provision paragraph 1.2 of the contract, to the appointment of [*insert name*] as site manager from [*insert date*].

[*If replacement:*]

The present site manager, [*insert name*], will be leaving on [*insert date*] and we should be pleased to receive your written consent, as required by Schedule 2, supplemental provision paragraph 1.2 of the contract, to the appointment of [*insert name*] as site manager from that date.

Yours faithfully

Letter 53
To architect, if required to furnish names and addresses of operatives
This letter is only suitable for use with GC/Works/1 (1998).

Dear

Thank you for your letter of the [*insert date*].

In accordance with clause 26(2) of the conditions of contract, we have pleasure in enclosing a full list of the names and addresses of persons who are concerned with the works and the capacities in which they are so concerned. Please inform us if you require further particulars. They are all persons whom we employ and for whom we are responsible.

There may well be other persons not in our employ who are concerned with the works for whom we do not have, nor do we have the authority to obtain, such information.

Yours faithfully

Letter 54
To architect, if passes are required
This letter is only suitable for use with GC/Works/1 (1998).

Dear

In accordance with clause 27 of the conditions of contract, we enclose a list of persons requiring passes to secure admission to the site. Please inform us what, if any, other details you require in respect of each person.

Yours faithfully

Letter 55
To employer, regarding the employer's representative
This letter is only suitable for use with SBC.

Dear

[*Either:*]

We note that M.. [*insert name*] has been appointed as the employer's representative under clause 3.3 of the contract.

[*Or:*]

We note that M.. [*insert name*] has been appointed as project manager. As there is no provision for such a role in the contract, we assume that he/she [*delete as appropriate*] has been appointed as the employer's representative under clause 3.3 of the contract.

[*Then:*]

We should be grateful if you would confirm, as required by clause 3.3, that M.. [*insert name*] will exercise all the functions ascribed to you in the contract and specify what if any exceptions there are.

Yours faithfully

Letter 56
To employer, regarding the project manager
This letter is not suitable for use with SBC, DB OR GC/Works/1 (1998).

Dear

We write simply in order to clarify something. We note that you have appointed a project manager. The contract makes no provision for such a person, the key people in the contract being the employer, the contractor, the architect and the quantity surveyor. It is obvious, if the project is to proceed smoothly, that everyone should be aware of his or her role and the powers and duties that go with it.

We shall, therefore, assume that the project manager is your own representative and we should be grateful if you would advise him/her [*delete as appropriate*] of the situation bearing in mind that only the architect may issue instructions, certificates and extensions of time. If the project manager attends site meetings, he/she [*delete as appropriate*] should not become involved unless invited to do so by the architect or by us. We trust that you will agree the importance of getting these points clear at the commencement of the project if serious confusion is to be avoided.

If you disagree with our view of the situation, we should be grateful to receive your own views as soon as possible.

Yours faithfully

Copy: Architect

4   Operations on Site

Unless the project is very small, a great many letters will be written during this stage. Some sizeable matters, such as payment, extensions and loss and / or expense, have been dealt with in separate sections to make location easier, but there are a number of other matters which can be the subject of standard letters.

Important letters may be written about the master programme, possession of the site, drawings, site meetings, instructions, defective materials and antiquities. DB generates certain letters not found in connection with the traditional contract. Such topics as submission and comments on drawings, discrepancies in or between Employer's Requirements and Contractor's Proposals and changes in statutory requirements or development control requirements are included.

Contractors sometimes have problems with the clerk of works and it is often difficult to know what steps to take because the clerk of works usually enjoys the support of the architect. There is no reason why the contractor should have to put up with some irregular practices and a few suggested letters for regrettably common situations are included.

Most letters written during this period, of course, will not be standard. They will be written in response to particular circumstances. Employing standard letters for fairly routine matters will free some time for dealing with a non-standard situation.

Letter 57
To employer, if possession not given on the due or the deferred date (a)
This letter is not suitable for use with MW or MWD or GC/Works/1 (1998).
Special delivery

Dear

[*If deferment of possession clause does not apply:*]

Possession of the site should have been given to us on [*insert date*] in accordance with clause 2.4 [*substitute '2.3' when using DB*] of the conditions of contract. Possession was not given to us on the due date.

[*If deferment of possession clause does apply, but deferred date not met:*]

Under clause 2.5 [*substitute clause '2.4' when using DB*] you deferred the giving of possession until [*insert date*]. You did not give us possession on the deferred date.

[*Then either:*]

Moreover, the architect has informed us today by telephone that the date for possession remains uncertain.

[*Or:*]

The architect has informed us today by telephone that you will be unable to give possession until [*insert date*].

[*continued*]

Letter 57 continued

[*Then:*]

By copy of this letter to the architect, we give notice of delay under clause 2.27.1 [*substitute '2.19.1' when using IC or ICD or 2.24.1' when using DB*] involving a Relevant Event under clause 2.29.6 [*substitute '2.20.6' when using IC or ICD or '2.26.5' when using DB*] and make application for loss and/or expense under clause 4.23 [*substitute '4.17' when using IC or ICD or '4.19' when using DB*] involving a Relevant Matter under clause 4.24.5 [*substitute '4.18.5' when using IC or ICD or '4.20.5' when using DB*]. However, such notice and application is without prejudice to any other rights and remedies including (but without limitation) the right to treat the failure to give possession as a serious breach of contract entitling us to damages and, if prolonged, to treat it as a repudiation of the contract on your part.

Naturally, we hope that your difficulties will be resolved and we should welcome a meeting with you to discuss the prospects of taking possession of the site. To that end we are generally available on [*insert list of dates*] and suggest that you telephone us to arrange a meeting.

Yours faithfully

Copy: Architect

Letter 58
To employer, if possession not given on the due date (b)
This letter is only suitable for use with MW and MWD.
Special delivery

Dear

Possession of the site should have been given to enable us to commence the works on [*insert date*] in accordance with clause 2.2 [*substitute '2.3' when using MWD*] of the conditions of contract. Possession was not given to us on the due date.

[*Then either:*]

The architect has informed us today by telephone that the date when you will be able to give possession is still uncertain.

[*Or:*]

The architect has informed us today by telephone that you will be unable to give possession until [*insert date*].

[*Then:*]

This is a serious breach of contract for which we will require appropriate compensation and we reserve all our rights and remedies in this matter. Without prejudice to the foregoing, we suggest that a meeting would be useful and look forward to hearing from you.

Yours faithfully

Letter 59
To architect, if date for possession advanced
This letter is not suitable for use with GC/Works/1 (1998).

Dear

Thank you for your letter of the [*insert date*] advising us that the employer is prepared to allow us to take possession of the site on [*insert date*].

Naturally, we will endeavour to take advantage of the opportunity to make an earlier start than we anticipated and we will inform you when we intend to take possession. There are many things to take into consideration and some reorganising of our labour resources would be involved, all of which would result in additional costs, which we would expect the employer to reimburse. Please let us know, by return, if the employer will reimburse such costs so that we can make our decision accordingly. If we take possession of the site on the date you suggest, such action will not affect our obligation to complete the works on the date for completion stated in the Contract Particulars.

Yours faithfully

Letter 60

To employer, giving consent to the engagement of other persons
This letter is not suitable for use with MW or MWD or GC/Works/1 (1998).

Dear

We are in receipt of your letter/an architect's instruction [*delete as appropriate and insert date*] referring to the engagement of [*insert name*] in accordance with clause 2.7.2 [*substitute '2.6.2' when using DB*] of the conditions of contract. The work is not detailed in the contract bills [*or specification or Employer's Requirements*], but we understand that it will consist of [*insert details*] and it will commence on site on the [*insert date*].

We will give our consent to the work if you will confirm in writing that you recognise our entitlement to extension of time and loss and/or expense in consequence.

Yours faithfully

Copy: Architect

Letter 61
To employer, withholding consent to the engagement of other persons
This letter is not suitable for use with MW or MWD or GC/Works/1 (1998).

Dear

We are in receipt of your letter/an architect's instruction [*delete as appropriate and insert date*] from which we see that it is intended to engage [*insert name*] in accordance with clause 2.7.2 [*substitute '2.6.2' when using DB*] of the conditions of contract. The work is not detailed in the contract bills [*or specification*], but we understand that it will consist of [*insert details*] and it is proposed to commence on site on [*insert date*] reaching completion on the [*insert date*].

We are unable to give our consent to the work, because [*insert details of reasonable objection*].

Yours faithfully

Copy: Architect

Letter 62
To architect, regarding items in minutes of site meeting

Dear

We have examined the minutes of the meeting held on the [*insert date*], which we received today. We have the following comments to make:

[*List comments.*]

Please arrange to have these comments published at the next meeting and inserted in the appropriate place in the minutes.

Yours faithfully

Copies: All present at meeting and those included in the original circulation

Letter 63
To architect, enclosing master programme (a)
This letter is only suitable for use with SBC.

Dear

In accordance with clause 2.9.1.2 of the conditions of contract, we have pleasure in enclosing two copies of our master programme.

We should be pleased to receive your approval as soon as possible so that we can proceed with project planning.

Yours faithfully

Letter 64
To architect, enclosing master programme (b)
This letter is not suitable for use with SBC.

Dear

We have pleasure in enclosing two copies of our master programme. Please let us have your approval as soon as possible so that we can proceed with project planning.

Yours faithfully

Letter 65
To architect, enclosing revision to the master programme
This letter is only suitable for use with SBC.

Dear

In accordance with clause 2.1.9.2 of the conditions of contract, we have pleasure in enclosing two copies of our master programme revision [*insert revision number*] to take account of your decision under clause 2.28/ confirmation of a Schedule 2 quotation [*delete as appropriate*].

We should be pleased to have your approval to the revised programme.

Yours faithfully

Letter 66
To architect, requesting information

Dear

We should be pleased to receive the following information which it is necessary for us to receive by the dates stated in order to enable us to carry out and complete the Works in accordance with the conditions of contract.

[*List the information required and the date by which each item of information must be received. Wherever possible, allow the architect at least 14 days to prepare the information.*]

Yours faithfully

Letter 67
To architect, if insufficient information on setting out drawings
This letter is not suitable for use with DB.

Dear

We are preparing to commence work on site on [*insert date*]. Our first task will be to set out the Works. An examination of the drawings you have supplied to us reveals that there is insufficient information for us to set out the Works accurately. We enclose a copy of your drawing number [*insert number*] on which we have indicated in red the positions where we need dimensions/levels [*delete as appropriate*].

We need this information by [*insert date*] in order to avoid delay and disruption to the Works.

Yours faithfully

Letter 68
To architect, requesting information that setting out is correct
This letter is not suitable for use with DB.

Dear

We refer to our letter of the [*insert date*] in which we informed you that the information on your drawings was insufficient to enable us to set out the Works accurately. You responded by telephone, asking us to set out the Works to the best of our ability based on the information provided. We have carried out your instructions and we should be pleased to receive your confirmation that the setting out is correct. If we do not receive such confirmation, in writing, by return of post, we shall be obliged to notify you of delay to the Works and disruption for which we will seek appropriate financial recompense.

Yours faithfully

Letter 69
To architect, if information received late
This letter is only suitable for use with SBC, IC and ICD.

Dear

On [*insert date*] we requested the following drawings/details/instructions [*delete as appropriate*] by [*insert date*]. They have not been received.

Clause 2.12.1 [*substitute '2.11.1' when using IC or ICD*] requires you to provide such information 'as and when from time to time may be necessary'. You are clearly in breach of your obligations. We have complied with our duty in the last part of the clause, to advise you sufficiently in advance of the time when it is necessary for us to receive the information. The absence of the information is causing us delay and disruption. Please inform us, by return, when we can expect it to arrive.

Take this as notice of delay under clause 2.27.1 [*substitute '2.19.1' when using IC or ICD*] and an application for loss and/or expense under clause 4.23 [*substitute '4.17' when using IC or ICD*]. We expect to be able to provide further details in due course.

Yours faithfully

Letter 70
To architect, if information not received in accordance with the information release schedule
This letter is only suitable for use with SBC, IC and ICD.

Dear

We draw your attention to the information release schedule for this project which indicates that [*describe the information and the date on which it should have been provided*]. It has not been received.

You are clearly in breach of your obligations as set out in clause 2.11 [*substitute '2.10' when using IC or '2.10.1' when using ICD*]. The absence of the information is causing us delay and disruption. Please inform us, by return, when we can expect it to arrive.

Take this as notice of delay under clause 2.27.1 [*substitute '2.19.1' when using IC or ICD*] and an application for loss and/or expense under clause 4.23 [*substitute '4.17' when using IC or ICD*]. We expect to be able to provide further details in due course.

Yours faithfully

Letter 71
To architect, if design fault in architect's or consultant's drawings
This letter is not suitable for use with DB.

Dear

We should be pleased if you would give further consideration to your drawing number [*insert number*]. In our view [*describe the problem*]. Therefore, if we proceed on the basis of this drawing without amendment, it is likely/possible [*delete as appropriate*] that serious defects or damage will result for which we can accept no liability.

Please let us have your comments and amended drawing by [*insert date*] to avoid delay and disruption to the progress of the Works.

Yours faithfully

Letter 72
To architect, if a design fault in architect's or consultant's drawings which the architect refuses to correct
This letter is not suitable for use with DB.
Special/recorded delivery

Dear

We refer to our letter of the [*insert date*] warning you of a design fault in your drawing number [*insert number*]. We note that, by your letter of the [*insert date*], you decline to amend your design to take account of our comments.

[*Either:*]

We have discharged any duty we may have to warn you and the employer of known design defects. Please take this as formal notice that we will carry out your instructions to proceed to construct in strict accordance with the above mentioned drawing, but we will not accept any liability for subsequent failure.

[*Or, if the design fault could give rise to injury or death of persons:*]

It is clear that if we were to proceed to construct the design fault in question, a serious injury or even death could result. In these circumstances it would be irresponsible of this company to proceed with the construction. We, therefore, intend to seek immediate adjudication on the question and we trust that you will co-operate in obtaining a quick decision. In due course, when matters are clearer, we shall be notifying delay and making application for loss and/or expense as a result of the delay and disruption.

Yours faithfully

Copy: Employer

Letter 73
To architect, if contractor providing contractor's design documents
This letter is only suitable for use with the contractor's designed portion of SBC and with DB.
Special delivery

Dear

In accordance with the submission procedure in Schedule 1, we enclose two copies of the following contractor's design documents:

[*List drawings, specifications, details and, if requested, calculations.*]

Please return the documents within 14 days of receipt, marked 'A', 'B' or 'C'. Please note that where you return a document marked 'B' or 'C', paragraph 4 requires you to identify in writing why you consider it is not in accordance with the contract.

Any documents not returned within 14 days will be regarded as marked 'A' in accordance with paragraph 3.

Yours faithfully

Letter 74
To architect, who fails to return the contractor's drawings in due time
This letter is only suitable for use with SBC or DB.

Dear

We refer to our letter of the [*insert date*] enclosing two copies of drawings numbered [*insert numbers*], being the contractor's design documents that we submitted under paragraph 1 of the Schedule 1 design submission procedure. Under paragraph 2, you should have responded within 14 days of receipt of our submission. At the date of this letter, we have not received such response.

Paragraph 3 provides that, in these circumstances, the design documents will be regarded as marked 'A' and we are proceeding accordingly. Please take note that if you subsequently return the documents with comments attached, we will quite properly consider them to be instructions requiring variations to the Works and, as well as the cost of such variations, we shall be entitled to extension of time and loss and/or expense.

Yours faithfully

Letter 75
To architect, if architect returns contractor's design document marked 'C'
This letter is only suitable for use with the contractor's designed portion of SBC or with DB.
Special delivery

Dear

[*Either:*]

We have pleasure in enclosing drawings numbers [*insert numbers*] which we have amended in accordance with your comments. We should be pleased if you would return them, marked 'A' by close of business on [*insert date*] so that the Works are not delayed.

[*Or:*]

We are in receipt of drawings numbers [*insert numbers*], which were marked 'C' and which contained your comments stating why you considered that they were not in accordance with the contract. We disagree with your comments as set out in more detail in the statement below and we consider that compliance with your comments will give rise to variations. We look forward to receiving your confirmation of instruction requiring a variation in each instance.

[*Set out the numbers of the drawings in dispute and, against each one, state why the drawing is in accordance with the contract and why the comment is wrong.*]

Yours faithfully

Letter 76
To architect, if architect confirms a comment on documents marked 'C'
This letter is only suitable for use with the contractor's designed portion of SBC or with DB.
Actual/special delivery

Dear

We refer to your letter dated [*insert date*] in answer to ours of the [*insert date*] where we considered that certain drawings were in accordance with the contract and that your comments amounted to variations. We note that you now confirm your comments on drawings numbers [*insert numbers*]. We, therefore, enclose drawings numbers [*insert numbers*] which we have amended in accordance with your comments. We should be pleased if you would return them, marked 'A' by close of business on [*insert date*] so that the Works are not delayed.

Notwithstanding the provisions of paragraph 8.1 of Schedule 1, we still consider your amendments to be variations and we expect your confirmation of the same by return.

Without prejudice to the above, we consider your comments to be an instruction. Take this as confirmation of such instruction under clause 3.12.2 [*substitute '3.7.2' when using DB*].

Yours faithfully

Letter 77

To architect, if contractor providing contractor's design documents
This letter is only suitable for use with the contractor's designed portion of ICD or MWD.
Actual/special delivery

Dear

In accordance with clause 2.10.2.1 [*substitute '2.1.5' when using MWD*] of the conditions of contract, we enclose two copies of the following contractor's design documents:

[*List drawings, specifications, details and, if requested, calculations.*]

[*If there is a submission procedure stated in the contract, follow it, if there is no procedure, add:*]

Please confirm within 7 days of receipt that we may proceed with construction using these documents. Bear in mind that if you instruct us to amend any of the documents, such amendment will be a variation to be valued under section 5 [*substitute 'clause 3.6' when using MWD*] of the contract unless such documents can be demonstrated to be not in accordance with the contract. If you do not respond within 7 days, we shall proceed on the basis that you have approved the documents.

Yours faithfully

Letter 78

To architect, if contractor providing levels and setting out information
This letter is only suitable for use with the contractor's designed portion of SBC, ICD or MWD or with DB.
Actual/special delivery

Dear

In accordance with clause 2.9.2.2 [*substitute '2.10.2.2' when using ICD or '2.1.5' when using MWD or '2.8' when using DB*] of the conditions of contract, we enclose two copies of the levels and setting out dimensions we propose to use for the Works.

In order to avoid any delays, please let us know by close of business on [insert the latest date you need to know, allowing for any changes that may need to be made after notification] if you have any adverse comments. If we do not receive any comments, we shall assume that you approve the information and proceed to use it in constructing the Works.

Yours faithfully

Letter 79
To architect, who returns contractor's drawings with comments
This letter is only suitable for use with ICD or MWD.

Dear

Thank you for your letter of the [*insert date*] with which you returned our drawings numbers [*insert numbers*] with comments.

Although we are always grateful for any comments you feel able to make, we should point out that the purpose of clause 2.10.2.1 [*substitute '2.1.5' when using MWD*] of the conditions of contract is to provide you with copies of the drawings and other information we intend to use to carry out the work. The drawings have been prepared strictly in accordance with the Employer's Requirements [*add, if using ICD: 'and our Contractor's Proposals'*].

If, therefore, you intend us to act upon your comments, please issue them as instructions requiring a variation under the provisions of clause 3.11.3 [*substitute '3.4' when using MWD*]. If we have no reasonable objection, such an instruction would be subject to valuation under clause 5.7 [*substitute '3.6' when using MWD*], extension of time under clause 2.19 [*substitute '2.8' when using MWD*] and loss and/or expense under clause 4.17 [*substitute '3.6.3' when using MWD*].

Yours faithfully

Letter 80
To architect, if discrepancy found between documents
This letter is not suitable for use with DB.

Dear

In accordance with clause 2.15 [*substitute '2.13.3' when using IC or ICD or '2.4'
when using MW or MWD or '2(3)' when using GC/Works/1(1998)*] we bring to
your attention the following discrepancies [*substitute 'inconsistencies' when
using IC, ICD, MW or MWD*] which we have discovered:

[*List, giving precise details of bills of quantities, specification or work schedule
references, drawing numbers or dates and numbers of architect's instructions.*]

In order to avoid delay or disruption to our progress, we require your
instructions by [*insert date*].

Yours faithfully

Letter 81

To architect, if discrepancy within the Employer's Requirements
This letter is only suitable for use with contractor's designed portion of SBC, ICD or MWD or with DB.

Dear

We have found a discrepancy in the Employer's Requirements [*give details*]. <u>Our Contractor's Proposals do not deal with the matter and, therefore,</u> [*delete the underlined portion when using MWD*] we propose the following amendment:

[*Describe in detail how to deal with the discrepancy including the provision of any drawings.*]

Please either let us have your written agreement to our proposal or details of your alternative decision, either of which will rank as a change to the Employer's Requirements. We need your agreement or decision by [*insert date*] in order to avoid delay to the Works.

Yours faithfully

Letter 82

To architect, if discrepancy within the Contractor's Proposals
This letter is only suitable for use with contractor's designed portion of SBC or ICD or DB.

Dear

We have found a discrepancy in our Contractor's Proposals [*give details*]. We propose the following amendment to remove the discrepancy:

[*Describe in detail how to deal with the discrepancy including the provision of any drawings.*]

In accordance with clause 2.16.1 [*substitute '2.13.1.4' when using ICD or '2.14.1' when using DB*] of the conditions of contract, please either decide which of the discrepant items you prefer or accept our proposed amendment. We require your decision or acceptance in writing by [*insert date*] in order to avoid delay to the Works.

Yours faithfully

Letter 83
To architect, if discrepancy found between Employer's Requirements and Contractor's Proposals
This letter is only suitable for use with contractor's designed portion of SBC or ICD or DB.

Dear

We have found a discrepancy between the Employer's Requirements and our Contractor's Proposals [*give details*]. The contract does not expressly deal with this situation. If you are content to accept the way we have dealt with the matter in our Contractor's Proposals, there is no problem, otherwise we suggest that a meeting on site is required. Please let us have your response as soon as possible/as a matter of urgency [*delete as appropriate*].

Yours faithfully

Letter 84
To architect, if alleging that contractor should have checked the design
This letter is only suitable for use with SBC, ICD, MWD or DB.

Dear

We are in receipt of your letter dated [*insert date*] in which you assert that we are responsible for the design of [*insert the element concerned*] or at the very least for checking it. This element is a design which you prepared and forms part of the Employer's Requirements.

Our obligation under clause 2.2 [*substitute '2.1' when using ICD, MWD or DB*] is to complete the design of the contractor's designed portion [*substitute 'design for the Works' when using DB*]. This duty is clarified by clause 2.13.2 [*substitute '2.34.4' when using ICD, '2.1.2' when using MWD or '2.11' when using DB*] which states that we are not responsible for the contents of the Employer's Requirements nor for verifying the accuracy of any design contained within them.

Therefore, we are neither responsible for the design to which you refer nor for checking that it works.

Yours faithfully

Letter 85
To employer, pointing out design error in Employer's Requirements
This letter is only suitable for use with SBC, ICD, MWD or DB.

Dear

In carrying out the detail design for which we are responsible, we have discovered that part of the design which you prepared and which forms part of the Employer's Requirements appears to be defective in that [*briefly indicate the nature of the defective design*].

Our obligation under clause 2.2 [*substitute '2.1' when using ICD, MWD or DB*] is merely to complete the design of the contractor's designed portion [*substitute 'design for the Works' when using DB*]. This duty is clarified by clause 2.13.2 [*substitute '2.34.4' when using ICD, '2.1.2' when using MWD or '2.11' when using DB*] which expressly states that we are not responsible for the contents of the Employer's Requirements nor for verifying the accuracy of any design contained within them.

We believe that it is now a matter for you to issue a change instruction to vary the Employer's Requirements to enable us to proceed with our part of the design. Please be aware that this is now causing us a delay for which we shall expect an appropriate extension of time in due course. It may also give rise to direct loss and/or expense for which we will seek reimbursement as soon as the position becomes clear.

Yours faithfully

Letter 86
To architect, requesting directions to integrate the design
This letter is only suitable for use with SBC, ICD or MWD.

Dear

We refer to [*insert description of the element concerned*], which is part of the contractor's designed portion of the Works.

It is unclear how this element is to be integrated into the design of the Works as a whole. [*Add either:*] This was not clear in the invitation to tender documents. [*Or:*] This is due to the issue of your instruction number [*insert number*] dated [*insert date*].

Therefore, we request your instructions for integration under clause 2.2.2 [*substitute '2.1.2' when using ICD or '2.1.3' when using MWD*] by [*insert date*] so as to avoid delay and extra costs to the project.

Yours faithfully

Letter 87
To architect, noting divergence between statutory requirements and other documents (a)
This letter is not suitable for use with DB.

Dear

We have found what appears to be a divergence between statutory requirements and your drawing/detail/schedule/bills of quantities/specification [*delete as appropriate and add number of drawing, page and item number of bills of quantities etc.*] as follows:

[*Insert details of the divergence.*]

Please let us have your instruction by [*insert date – note that SBC, IC and ICD allow the architect 7 days from receipt of the notice to respond with an instruction. So far as MW and MWD are concerned, insert a reasonable date.*]. Failure to issue your instruction by that date will result in delay and disruption to the Works for which we will seek appropriate financial recompense.

Yours faithfully

Letter 88
To architect, noting divergence between statutory requirements and other documents (b)
This letter is only suitable for use with DB.

Dear

We have found what appears to be a divergence between statutory requirements and the Employer's Requirements/our Contractor's Proposals [*delete as appropriate*] as follows:

[*Insert details of the divergence.*]

We propose the following amendment to remove the divergence:

[*Describe in detail how to deal with the divergence including the provision of any drawing.*]

Please let us have your written consent to our proposal by [*insert date*] in order to avoid delay to the Works. Please note the amendment on the contract documents in accordance with clause 2.15.1 of the conditions of contract and send us a copy.

Yours faithfully

Letter 89
To architect, if emergency compliance with statutory requirements required
This letter is not suitable for use with MW, MWD or GC/Works/1 (1998).

Dear

We hereby give notice as required by clause 2.18.2 [*substitute '2.16.2' when using IC, ICD or DB*] of the conditions of contract that we have been obliged to carry out work constituting emergency compliance with statutory requirements. The emergency and the steps we are taking/have taken [*delete as appropriate*] are as follows:

[*Insert details.*]

We confirm that the work carried out and materials supplied rank as a variation [*substitute 'change' when using DB*] and we should be pleased to receive any further instructions necessary.

Please take this letter as a notice of delay under clause 2.27.1 [*substitute '2.19.1' when using IC or ICD or '2.24.1' when using DB*] and an application for loss and/or expense under clause 4.23 [*substitute '4.17' when using IC or ICD or '4.19' when using DB*].

Yours faithfully

Letter 90

To architect, if a change in statutory requirements after base date
This letter is not suitable for use with IC, MW, MWD or GC/Works/1 (1998).

Dear

On [*insert date*] there was a change in statutory requirements which necessitates an alteration or modification to the Contractor's Designed Portion [*substitute 'to the Works' when using DB*] [*give details*].

We are putting the necessary amendment in hand and we will write to you again as soon as we are in a position to value the amendment which is to be treated as a variation [*substitute 'change' when using DB*] instruction under the provisions of clause 2.17.2.1 [*substitute '2.15.2.1' when using ICD or DB*] of the conditions of contract. [*When using DB, add:*] In our view, this is not an instruction to be dealt with under supplemental provision 4.

[*Then add:*]

Therefore, please take this letter as notice of delay under clause 2.27.1 [*substitute '2.19.1' when using ICD or '2.24.1' when using DB*] and application for loss and/or expense under clause 4.23 [*substitute '4.17' when using ICD or '4.19' when using DB*].

Yours faithfully

Letter 91
To architect, if development control decision after base date
This letter is only suitable for use with DB.

Dear

We have just received a permission/approval [*delete as appropriate*] from [*state the relevant authority, e.g. local planning authority*] and a copy is enclosed. To make them conform, it will be necessary to amend our Contractor's Proposals as follows: [*give details*].

We are putting the necessary amendment in hand and we will write to you again as soon as we are in a position to value the amendment which is to be treated as a change instruction under the provisions of clause 2.15.2.2 of the conditions of contract. In our view, this is not an instruction to be dealt with under supplemental provision 4. Therefore, please take this letter as notice of delay under clause 2.24.1 and application for loss and/or expense under clause 4.19.

Yours faithfully

Letter 92
To employer (not being a local authority), objecting to the nomination of a replacement architect
This letter is not suitable for use with DB.

Dear

Under the provisions of clause 3.5.1 [*substitute '3.4.1' when using IC or ICD, 'article 3' when using MW or MWD or '1(1)' when using GC/Works/1 (1998)*] we hereby formally give notice of our objection to the nomination of [*insert name*] of [*insert address*] as architect [*substitute 'PM' when using GC/Works/1 (1998)*] for the purpose of this contract in succession to [*insert name and address of previous architect/PM*].

The grounds for our objections are [*insert particular reasons for objection*].

A good working relationship between architect [*substitute 'PM' when using GC/Works/1 (1998)*] and contractor is vital to the successful completion of any project. With this in mind, we look forward to hearing that you have reconsidered the nomination.

Yours faithfully

Letter 93

To employer (not being a local authority), objecting to the nomination of the employer as replacement architect
This letter is not suitable for use with DB or GC/Works/1 (1998).
By fax and special delivery

Dear

We were surprised to learn that you proposed to act as architect yourself as a replacement for the previous architect. Presumably, you are purporting to act under the provisions of clause 3.5.1 [*substitute '3.4.1' when using IC or ICD or 'article 3' when using MW or MWD*]. We are not prepared to accept you acting in this role. Not only are you not a person registered with the Architects Registration Board and, therefore, not permitted to use the title 'architect', we are advised that this kind of provision in a construction contract does not empower you to appoint yourself as architect and we suggest that you take your own urgent legal advice to confirm that position.

Unless you inform us by close of business on [*insert a date three working days from the date of this letter*] that you have reconsidered the nomination, we shall have no alternative but to seek adjudication on the point and on the damage, loss and expense caused to us by the breach.

Yours faithfully

Letter 94
To employer, if replacement architect not appointed
This letter is not suitable for use with DB or GC/Works/1 (1998).
By fax and post

Dear

We were notified/became aware [*delete as appropriate*] that the architect named in the contract had ceased to act on or about the [*insert date*]. Under the provisions of article 3 you are obliged to nominate a replacement architect within 21 [*substitute '14' when using IC, ICD, MW or MWD*] days.

It is now some [*insert number*] days since you should have made the appointment and you have failed to do so. The result of this is that, at the most basic level, no instructions or certificates can be issued and the extension of time provisions cannot be operated. Therefore, as soon as we require an instruction to enable us to proceed, we shall be obliged to stop that particular activity although we will, for the record, notify you of the instruction required. We envisage that the whole site will be at a standstill very shortly and we must consider our position. In any event, time will soon become at large.

Any replacement architect will need time to absorb the detail of this project and we urge you to make the appointment immediately. In the meantime you are liable to us for all the damage, loss and expense caused to us by your breach.

Yours faithfully

Letter 95
To employer, objecting to the nomination of a replacement quantity surveyor
This letter is only suitable for use with SBC, IC or ICD.

Dear

Under the provisions of clause 3.5.1 [*substitute '3.4.1' when using IC or ICD*] we hereby formally give notice of our objection to the nomination of [*insert name*] of [*insert address*] as quantity surveyor for the purpose of this contract in succession to [*insert name and address of previous quantity surveyor*].

The grounds for our objection are [*insert particular reasons for objection*].

A good working relationship between quantity surveyor and contractor is vital to the successful completion of any project. With this in mind, we look forward to hearing that you have reconsidered the nomination.

Yours faithfully

Letter 96
To architect, regarding directions issued on site by the clerk of works
This letter is not suitable for use with GC/Works/1 (1998).

Dear

The clerk of works has issued direction number [*insert number*] dated [*insert date*] on site, a copy of which is enclosed.

Such directions have, of course, no contractual effect [*insert 'unless you confirm them within 2 working days of issue' when using SBC*]. Clearly, the directions of the clerk of works issued in relation to the correction of defective work can be very helpful. We are anxious to avoid misunderstandings on site and in this spirit we suggest that the clerk of works should issue no further directions, other than those relating to defective work. All other matters can be referred directly to you by telephone and, at your discretion, a proper architect's instruction can be issued.

In our view, the above system would remove a good deal of the uncertainty which must result from the present state of affairs. We look forward to hearing your comments.

Yours faithfully

Letter 97
To architect, regarding instructions issued on site by the clerk of works
This letter is only suitable for use with GC/Works/1 (1998).

Dear

The clerk of works has issued instruction number [*insert number*] dated [*insert date*] on site, a copy of which is enclosed.

Clause 4(2) of the conditions of contract provides that the clerk of works may issue only instructions under clause 31 unless you expressly delegate other powers to the clerk of works in writing and notify us under the provisions of clause 4(4).

The enclosed instruction does not fall under the provisions of clause 31 and we have not yet received a clause 4(4) notice from you. Our agent on site has strict instructions from us that instructions of the clerk of works are not to be executed unless expressly empowered by the contract. If you wish the instruction to be carried out, please let us have your confirmation without delay or, alternatively, a notice under clause 4(4) that the clerk of works has power to issue any instructions which the contract empowers you to issue.

Yours faithfully

Letter 98
To architect, if clerk of works defaces work or materials

Dear

It is common practice for the clerk of works to deface work or materials which are considered to be defective. We assume that the basis for such action is to bring the defect to the notice of the contractor and ensure that it cannot remain without attention.

We object to the practice on the following grounds:

1. The work or materials so marked may not be defective and we will be involved in extra work and the employer in extra costs in such circumstances, because otherwise satisfactory work or materials will have been spoilt.
2. The work or materials so marked, if indeed defective, will not be paid for and will be our property when removed. We may be able to incorporate it in other projects where a different standard is required. Defacement by the clerk of works would prevent such re-use.

We will take no point about the defacing marks we noted on site today, but if the practice continues, we will seek financial reimbursement on every occasion.

Yours faithfully

Letter 99
To architect, if numerous 'specialist' clerks of works visiting site
This letter is not suitable for use with DB, MW or MWD.

Dear

Clause 3.4 [*substitute '3.3' when using IC or ICD or '4(2)' when using GC/Works/1 (1998)*] of the conditions of contract permits the employer to appoint a clerk of works. Your letter of the [*insert date*] informed us that the clerk of works would be [*insert name*]. During the past two weeks a number of persons have presented themselves at our site office purporting to be 'specialist clerks of works' [*or substitute the actual title*] and demanding access to the Works.

These persons are neither known to us nor included in the contract and, therefore, we have exercised our right to exclude them from the Works.

We already permit inspections by those consultants introduced at the project start meeting of the [*insert date*] even though the contract is silent as to their existence, because we desire to extend all reasonable co-operation. To allow 'specialist clerks of works' free access would not, in our view, be reasonable. If you require us to grant them such access, please so inform us in writing, but note that, in such circumstances, we consider that our progress would be hindered and we will take appropriate legal advice in regard to our rights and remedies.

Yours faithfully

Letter 100
To architect, if clerk of works instructs operatives direct

Dear

May we draw your attention to an unfortunate situation which is developing on site? The clerk of works, with whom we have had the most cordial relations, is getting into the habit of giving oral directions to our operatives on site even going so far as to reprimand some of them for poor workmanship.

We very much value all the comments of the clerk of works, provided that they are addressed to the person-in-charge/site manager [*delete as appropriate*]. The present situation is causing, as one might expect, a degree of disruption as we have to waste valuable time smoothing ruffled feathers. We have spoken to the clerk of works informally, but to no effect. We are anxious to avoid tension on the site, which would be in no one's best interests and we should be grateful if you would deal with this matter tactfully as soon as possible.

Yours faithfully

Letter 101
To quantity surveyor, submitting a Schedule 2 quotation
This letter is only suitable for use with SBC.

Dear

Under the provisions of Schedule 2, paragraph 1.2 and in compliance with architect's instruction number [*insert number*] dated [*insert date*], we enclose our quotation. The quotation comprises the required adjustment to the contract sum with calculations, the required adjustment to the time for completion of the Works, the amount of loss and/or expense required in lieu of any adjustment under clause 4.23 and the cost of preparing the quotation.

[*If the instruction specifically requires it, add:*]

Indicative information is included about the extra resources needed and the method of carrying out the work.

[*Then add:*]

The quotation is open for acceptance for 7 days from the date of receipt.

Yours faithfully

Letter 102
To architect, regarding verification of vouchers for daywork
This letter is not suitable for use with MW, MWD or GC/Works/1 (1998).

Dear

We enclose vouchers numbers [*insert numbers*] in respect of [*insert description of work*]. We should be pleased if you would verify them as required by clause 5.7 [*substitute '5.4' when using IC or ICD or '5.5' when using DB*] of the conditions of contract.

[*If appropriate, add:*]

The clerk of works always asks to see such vouchers and it occurred to us that it would save time if you notified us that the clerk of works was your authorised representative for the purpose of clause 5.7 [*substitute '5.4' when using IC or ICD or '5.5' when using DB*]. The clerk of works could officially verify the vouchers as required by the contract. We should be interested to hear whether this suggestion meets with your approval.

Yours faithfully

Letter 103
To employer, if disagreement over whether work is a variation or included in the contract
Special delivery

WITHOUT PREJUDICE

Dear

We refer to your letters of the [*insert dates*] and ours of the [*insert dates*] relating to [*describe work*]. It is our firm view that this work is not included in the contract and, therefore, constitutes a variation for which we are entitled to payment.

We are advised that we can refuse performance of the disputed work until you authorise a variation. If you continue to refuse to so authorise, we may be entitled to treat the contract as repudiated and sue for damages.

Without prejudice to our rights, we are prepared to carry out the work, leaving this matter in abeyance for future determination by adjudication or arbitration, if you will agree in writing and confirm that you will not deny our entitlement to payment in such reference if, on the true construction of the contract, the work is held to be not included.

Yours faithfully

Copy: Architect

Letter 104
To architect, requiring the specification of the clause empowering an instruction (a)
This letter is not suitable for use with MW, MWD or GC/Works/1 (1998).

Dear

We have received today your instruction number [*insert number*], dated [*insert date*] requiring us to [*insert substance of instruction*].

We request you, in accordance with clause 3.13 [*substitute '3.10' when using IC or ICD or '3.8' when using DB*] of the conditions of contract, to specify in writing the provision which empowers the issue of the above instruction.

Yours faithfully

Letter 105
To architect, requiring the specification of the clause empowering an instruction (b)
This letter is only suitable for use with MW, MWD or GC/Works/1 (1998).

Dear Sir

We have received today your instruction number [*insert number*], dated [*insert date*] requiring us to [*insert substance of instruction*].

Mindful that you are only empowered to issue instructions set out in the contract and that we would not be entitled necessarily to payment for instructions which are not so empowered, we request you to specify in writing the provision of the contract which empowers the issue of the above instruction.

Yours faithfully

Letter 106
To architect, confirming an oral instruction
This letter is only suitable for use with SBC or DB.

Dear

We hereby confirm that on [*insert date*], you orally instructed us to [*insert the instruction in detail*]. Under the provisions of clause 3.12.2 [*substitute '3.7.2' when using DB*] of the conditions of contract, you have 7 days in which to dissent in writing.

[*Or:*]

On [*insert date*], you orally instructed us to [*insert the instruction in detail*]. We complied with your instruction, as you are aware, but neglected to confirm your instruction as provided by clause 3.12.2 [*substitute '3.7' when using DB*] of the conditions of contract because [*insert reason*]. We should be pleased, therefore, if you would exercise your power under clause 3.12.4 [*substitute '3.7.4' when using DB*] and confirm the instruction in writing.

Yours faithfully

Letter 107
To architect, requesting confirmation of an oral instruction
This letter is only suitable for use with GC/Works/1.

Dear

On [*insert date*], you orally instructed us to [*insert the instruction in detail*]. In accordance with clause 40(3) of the conditions of contract, we should be pleased to receive your written confirmation.

Yours faithfully

Letter 108

To architect, if oral instruction not confirmed in writing
This letter is only suitable for use with MW or MWD.

Dear

On [*insert date*], you orally instructed us to [*insert the instruction in detail*]. We have carried out your instruction, as you are aware, but we have not yet received written confirmation as provided by clause 3.4 of the conditions of contract. We should be pleased, therefore, if you would send us your written confirmation immediately.

Yours faithfully

Letter 109
To architect, objecting to exclusion of person from the Works (a)
This letter is not suitable for use with DB.
By fax and first class post

Dear

We are in receipt of your instruction number [*insert number*] dated [*insert date*] which requires us to exclude [*insert name*] from the Works.

The instruction appears to have been issued under the provisions of clause 3.21 [*substitute '3.17' when using IC or ICD, '3.8' when using MW or MWD or '40(2)(h)' when using GC/Works/1 (1998)*] of the contract. It is clear that the instruction must not be issued unreasonably, but in this instance you have provided no/no adequate [*delete as appropriate*] explanation.

Your insistence on the removal of our staff amounts to interference in our organisation of the Works which is not acceptable. Moreover, it will result in delays and additional costs which we shall be entitled to recover from you.

Unless you provide a satisfactory explanation for your instruction by return fax, a dispute will exist which we may refer to adjudication.

Yours faithfully

Letter 110
To employer, objecting to exclusion of person from the Works (b)
This letter is only suitable for use with DB.
By fax and first class post

Dear

We are in receipt of your instruction number [*insert number*] dated [*insert date*] which requires us to exclude [*insert name*] from the Works.

Unlike the position under some other JCT contracts, there is no express power in this contract which allows you to exclude anyone from the Works. Clause 3.5 only requires us to comply with instructions which you are expressly empowered to give. Moreover, we do not believe that such power would be implied.

Your insistence on the removal of our staff amounts to interference in our organisation of the Works which is not acceptable. Moreover, it would result in delays and additional costs which we would be entitled to recover from you.

Please confirm by return fax that you are withdrawing this invalid instruction.

Yours faithfully

Letter 111
To architect, objecting to instruction varying obligations or restrictions
This letter is not suitable for use with MW, MWD or GC/Works/1 (1998).

Dear

Thank you for your instruction number [*insert number*], dated [*insert date*] which we received today. Your instruction requires a variation [*substitute 'change' when using DB*] within the meaning of clause 5.1.2.

We have reasonable objection to complying with your instruction in that [*insert grounds of objection*].

Our objection is submitted under the provisions of clause 3.10.1 [*substitute '3.8' when using IC or ICD or '3.5' when using DB*] and we should be pleased if you would withdraw or revise your instructions in the light of our comments.

[*Add, if appropriate:*]

Please let us have your reply by [*insert date*] in order to avoid the possibility of delay or disruption due to [*insert as appropriate*].

Yours faithfully

Letter 112
To architect, withholding consent to instruction altering the design of the Works
This letter is only suitable for use with MWD or DB.

Dear

Thank you for your instruction number [*insert number*] dated [*insert date*] which we received today. Your instruction affects the design of the contractor's designed portion Works [*substitute 'makes necessary an alteration or modification in the design of the Works' when using DB*]. Clause 3.4.2 [*substitute '3.9.1' when using DB*] provides that you may not issue any such instruction [*substitute 'affect any such change' when using DB*] without our consent.

On this occasion, we regret that we cannot give our consent. Although our consent is not to be unreasonably withheld, we believe our position to be reasonable. [*Set out the reasonable reasons for withholding consent.*]

If you think it will be useful, we should be happy to discuss this matter further in a face to face meeting.

Yours faithfully

Letter 113
To architect, if attempting to vary the Contractor's Proposals
This letter is only suitable for use with SBC, ICD or DB.

Dear

Thank you for your instruction number [*insert number*], dated [*insert date*]
which we received today. Your instruction has the effect of varying the
Contractor's Proposals.

May we draw to your attention that such instruction is not valid under the
contract, because clause 3.14.3 [*substitute '3.11.3' when using ICD or '5.1.1' when
using DB*] makes clear that an instruction requiring a variation [*substitute 'a
change' when using DB*] can only refer to the Employer's Requirements.
Therefore, it is an instruction which you have no power to issue and with
which we have no power or duty to comply.

If you wish to re-issue the instruction, referring to the Employer's
Requirements, we will of course comply forthwith. Alternatively, you
may wish to discuss the matter with us and we will endeavour to assist.

Yours faithfully

Letter 114
To architect, on receipt of 7 day notice requiring compliance with instruction (a)
This letter is not suitable for use with GC/Works/1 (1998).

Dear

We have today received your notice dated [*insert date*] which you purport to issue under the provisions of clause 3.11 [*substitute '3.9' when using IC or ICD, '3.5' when using MW or MWD or '3.6' when using DB*] of the conditions of contract.

[*Add either:*]

We will comply with your instruction number [*insert number*], dated [*insert date*] forthwith, but such compliance is without prejudice to, and reserving, any other rights and remedies which we may possess.

[*Or:*]

We consider that we have already complied with your instruction number [*insert number*], dated [*insert date*]. Any attempt by the employer to employ other persons and/or deduct from any monies due or to become due to us will be deemed to be a serious breach of contract for which we will seek appropriate remedies. Without prejudice to the foregoing, if you will immediately withdraw your notice requiring compliance, M.. [*insert name*] will be happy to meet you on site to sort out what appears to be an unfortunate misunderstanding.

[*continued*]

Letter 114 continued

[*Or:*]

It is not reasonably practicable to comply as you require within the period you specify because [*insert reasons*]. You may be assured that we have not forgotten our obligations in this matter and we intend to carry out your instruction number [*insert number*], dated [*insert date*] as soon as [*indicate operation*] is complete. In the light of this explanation, we should be pleased to hear, by return, that you withdraw your notice requiring compliance. If we do not have your reply by [*insert date*], we will immediately comply, but take this as notice that such immediate compliance will give grounds for substantial claims for extension of time and loss and/or expense.

Yours faithfully

Copy: Employer

Letter 115
To architect, on receipt of notice requiring compliance with instruction (b)
This letter is only suitable for use with GC/Works/1 (1998).

Dear

We have today received your notice dated [*insert date*] which you purport to issue under the provisions of clause 53 of the conditions of contract.

[*Add either:*]

We will comply with your instruction number [*insert number*], dated [*insert date*] forthwith, but such compliance is without prejudice to, and reserving, any other rights and remedies which we may possess.

[*Or:*]

We consider that we have already complied with your instruction number [*insert number*], dated [*insert date*]. Any attempt by the employer to provide labour and/or any things, or enter into a contract with others and/or recover alleged additional costs and expenses from us will be deemed to be a serious breach of contract for which we will seek appropriate remedies. Without prejudice to the foregoing, if you will immediately withdraw your notice requiring compliance, M.. [*insert name*] will be happy to meet you on site to sort out what appears to be an unfortunate misunderstanding.

[*continued*]

Letter 115 continued

[*Or:*]

It is not reasonably practicable to comply as you require within the period you specify because [*insert reasons*]. You may be assured that we have not forgotten our obligations in this matter and we intend to carry out your instruction number [*insert number*], dated [*insert date*] as soon as [*indicate operation*] is complete. In the light of this explanation, we should be pleased to hear, by return, that you withdraw your notice requiring compliance. If we do not have your reply by [*insert date*], we will indeed immediately comply, but take this as notice that such immediate compliance will give grounds for substantial claims for extension of time and loss and/or expense.

Yours faithfully

Copy: Employer

Letter 116
To architect, if instruction will affect the contractor's designed portion
This letter is only suitable for use with SBC and ICD.

Dear

Thank you for your instruction/direction [*delete as appropriate*] number [*insert number*] dated [*insert date*] which we received today. It is our view that the instruction/direction [*delete as appropriate*] will injuriously affect the efficacy of the contractor's designed portion by [*describe the way in which it will so affect the CDP*].

Under clause 3.10.3 [*substitute '3.8.2' when using ICD*] your instruction/direction [*delete as appropriate*] will not now take effect unless you confirm it in writing.

Note that, if you do so confirm, we have put you on notice regarding the injurious effect and we take no responsibility for the result.

Yours faithfully

Copy: Employer

Letter 117
To architect, withholding consent if instruction will affect the contractor's design
This letter is only suitable for use with MWD or DB.

Dear

Thank you for your instruction number [*insert number*] dated [*insert date*] which we received today. The instruction is or makes necessary an alteration or modification in the design of the CDP Works [*substitute simply 'the Works' when using DB*] by [*describe the way in which it will affect the CDP Works or the Works as appropriate*].

We formally give notice that in the circumstances we believe it reasonable to withhold our consent under clause 3.4.2 [*substitute '3.9.1' when using DB*]. Under the provisions of this clause your instruction will not now take effect.

Yours faithfully

Copy: Employer

Letter 118
To architect, removal of unfixed materials
This letter is not suitable for use with MW, MWD or GC/Works/1 (1998).

Dear

[*State quantity and nature of goods or materials*] are presently stored on site. It is our view that these materials should be stored at [*name place*] because [*state reason*]. We should be pleased to receive your written consent to the removal from site of these materials in accordance with the provisions of clause 2.24 [*substitute '2.17' when using IC or ICD or '2.21' when using DB*] of the conditions of contract.

Yours faithfully

Letter 119
To architect, if materials are not procurable
This letter is only suitable for use with SBC or DB.

Dear

We have been informed by our suppliers [*insert name*] that [*insert description of material*] is not procurable because [*insert reason*]. A copy of their letter dated [*insert date*] is enclosed for your information. We can obtain [*insert description of material*] which you may consider to be an alternative which meets your requirements at a cost of [*insert cost*]. We should be pleased to have your instructions by [*insert date*] because there is a danger that there will be delay to the progress of the Works.

Yours faithfully

Letter 120
To architect, if wishing to substitute materials or goods

Dear

We refer to [*insert a description of the material or goods*] which is included in the specification/bills of quantities/Employer's Requirements/Contractor's Proposals [*delete as appropriate*] at [*insert page and item number*].

It may be better not to use [*insert description*] because [*insert brief reasons*]. Instead, we suggest that we should use [*insert description of suggested material or goods*] and we anticipate there would be a resultant overall saving/additional cost of [*insert amount*].

We look forward to receiving your written consent [*add 'in accordance with clause 2.3.1' when using SBC or 'in accordance with clause 2.2.1' when using DB*].

Yours faithfully

Letter 121
To architect, after failure of work or materials or goods
This letter is only suitable for use with IC or ICD.

Dear

[*If the architect has notified failure, begin:*]

Thank you for your letter of the [*insert date*], notifying us [*briefly give details*].

[*Otherwise, begin:*]

We have to notify you that [*briefly give details*] was found not to be in accordance with the contract on [*insert date which must be not more than 7 days before the date of this letter*].

[*Then add:*]

In accordance with clause 3.15.1, we propose to [*insert action to be taken to ensure there are no similar failures*] to ensure that there is no similar failure in these areas/materials/goods [*delete as appropriate*]. We should be pleased to receive your immediate written approval of our proposals.

Yours faithfully

Letter 122
To architect, if contractor objects to complying with
a clause 3.15.1 instruction
This letter is only suitable for use with IC or ICD.
Special/recorded delivery

Dear

We are in receipt of your instruction number [*insert number*] dated [*insert date*]
instructing us to [*insert nature of work*] which you purport to issue under clause
3.15.1 of the conditions of contract. We consider such instruction unreasonable
because [*state reasons*].

If within 7 days of receipt of this letter you do not in writing withdraw the
instruction or modify it to remove our objection, a dispute or difference will
exist as to whether the nature or extent of opening up/testing [*delete as
appropriate*] in your instruction is reasonable in all the circumstances. Such
dispute or difference will be referred to immediate adjudication. In such event,
we will comply with our obligations pending the result of such adjudication
and award of additional costs and extension of time.

Yours faithfully

Copy: Employer

Letter 123
To architect, after work opened up for inspection

Dear

We confirm that you inspected [*describe work*] opened up for your inspection in accordance with your instructions under clause 3.17 [*substitute '3.14' when using IC or ICD, '3.4' when using MW or MWD, '3.12' when using DB or '40(2)(i)' when using GC/Works/1 (1998)*] on [*insert date*] and found the materials, goods and work to be in accordance with the contract.

The cost of opening up and making good, therefore, is to be added to the contract sum and we will let you have details of our costs within the next few days. We will shortly send you details, particulars and estimate of the expected delay in completion of the works beyond the completion date and an application for reimbursement under the appropriate clause of the contract [*substitute 'reimbursement for consideration by the employer' when using MW or MWD*].

Yours faithfully

Letter 124
To architect, if excavations ready for inspection
This letter is only suitable for use with GC/Works/1 (1998).

Dear

The excavations for foundations for this project will be complete by [*insert date*].

In order to avoid delay, we require you, in accordance with clause 16, to examine such excavations immediately on completion and give us your written approval.

Yours faithfully

Letter 125
To architect, if issuing an instruction after ordering removal of defective work
This letter is only suitable for use with SBC or DB.

Dear

We are in receipt of your instruction number [*insert number*] dated [*insert date*] under clause 3.18.1 [*substitute '3.13.1' when using DB*] of the conditions of contract regarding the removal from site of work/materials/goods [*delete as appropriate*]. We have now received a further instruction number [*insert number*] dated [*insert date*] requiring a variation [*substitute 'change' when using DB*] in consequence under clause 3.18.3 [*substitute '3.13.2' when using DB*].

[*Then add either:*]

The work will be carried out as soon as reasonably practicable and we shall invite your approval when it is complete.

[*Or:*]

Your instruction is not reasonably necessary as a consequence of your clause 3.18.1 [*substitute '3.13.1' when using DB*] instruction because [*insert reasons*]. We request you to reconsider and to withdraw the instruction forthwith. If you inform us that you are not prepared to withdraw the instruction, we shall carry it out as soon as reasonably practicable, but without prejudice to our right, which we intend to exercise, to refer the matter to adjudication.

Yours faithfully

Letter 126

To architect, if issuing instruction for opening up after ordering removal of defective work
This letter is only suitable for use with SBC or DB.

Dear

We are in receipt of your instruction number [*insert number*] dated [*insert date*] under clause 3.18.1 [*substitute '3.13.1' when using DB*] of the conditions of contract regarding the removal from site of work/materials/goods [*delete as appropriate*]. We have now received a further instruction number [*insert number*] dated [*insert date*] requiring opening up for inspection under clause 3.18.4 [*substitute '3.13.3' when using DB*].

[*Then add either:*]

The work will be opened up as requested to be ready for your inspection on [*insert date*] at [*insert time*].

[*Or:*]

In our view, due regard has not been given to the code of practice in the contract and further, the instruction is not reasonable. We therefore propose to treat the instruction as being issued under the provisions of clause 3.17 [*substitute '3.12' when using DB*] and, if the work is found to be in accordance with the contract, we expect proper remuneration to include loss and/or expense and an appropriate extension of time. Please confirm your agreement by return or a dispute will have arisen between us and we shall carry out your instruction without prejudice to our right to refer the matter to adjudication or arbitration.

Yours faithfully

Letter 127
To architect, if issuing instruction for removal of defective work
This letter is only suitable for use with SBC.

Dear

We are in receipt of your instruction number [*insert number*] dated [*insert date*] instructing us to remove from site work which you state is not in accordance with the contract. You specify the work as [*repeat the architect's description*].

Our records show that the work in question was executed on [*insert date or dates*], some [*insert time period*] ago. The work was inherently a matter for the architect's satisfaction under clause 2.3. We draw your attention to clause 3.20 of the conditions of contract which stipulates that you must express any dissatisfaction with such work within a reasonable time from its execution. It is clear that, in breach of contract, you have not so expressed such dissatisfaction. We are now entitled to payment for complying with your instruction and we should be pleased to have your agreement by return.

Yours faithfully

Letter 128

To architect, if wrongly issuing instruction following failure to carry out the work in a proper and workmanlike manner
This letter is only suitable for use with SBC.

Dear

We are in receipt of your instruction number [*insert number*] dated [*insert date*] issued under clause 3.19 of the conditions of contract.

Such instruction is only to be issued following a failure to comply with clause 2.1 in regard to the carrying out of work in a proper and workmanlike manner or in accordance with the construction phase plan.

Our position is that we have complied with the requirements of the contract in this regard and although we shall, of course, comply with your instruction as if issued under clause 3.14, we shall require payment for doing so unless you can provide us with clear evidence of the failure to which you refer.

Yours faithfully

Letter 129
To architect, if work to be covered up
This letter is only suitable for use with GC/Works/1 (1998).

Dear

We hereby give notice, in accordance with clause 17 of the conditions of contract, that [*describe work*] will be covered up on [*insert date*]. After this date, any opening up required will be at the employer's expense.

Yours faithfully

Letter 130
To architect, if antiquities found
This letter is not suitable for use with IC, ICD, MW or MWD.

Dear

We have today uncovered [*describe*] which we consider falls within the provisions of clause 3.22 [*substitute '3.15' when using DB or '32(3)' when using GC/Works/1 (1998)*]. It lies at a depth of [*insert precise depth*] below [*insert some nearby permanent point from which measurement taken*] at [*describe location e.g. '2 metres west of proposed inspection chamber 23'*].

We have stopped work in the vicinity and erected a temporary wooden fence and waterproof cover. We should be pleased to receive your immediate written instructions.

[*If using SBC or DB, add:*]

The progress of the Works is being delayed by the above circumstances which we consider to be a relevant event under clause 2.29.2.1 [*substitute '2.26.2.1' when using DB*]. When the delay is finished we will furnish you with our estimate of delay in the completion of the Works and further supporting particulars. You may be assured that we are using our best endeavours to prevent delay in progress and completion of the Works. Please take this as notice in accordance with clause 2.27.1 [*substitute '2.24.1' when using DB*]. We further consider that we are entitled to reimbursement for direct loss and/or expense under the provisions of clause 3.24 [*substitute '3.17' when using DB*].

[*continued*]

Letter 130 continued

[If using GC/Works/1 (1998), add:]

In accordance with clause 36(1) we notify you that it is apparent that the Works will not be complete by the date for completion. We are unable to estimate the total delay at this stage, but when it is over, we will submit full details immediately. You may be assured that we are using our best endeavours to prevent delays and to minimise unavoidable delays.

The regular progress of the Works is likely to be materially disrupted and prolonged due to the above circumstances. In consequence, we will properly and directly incur expense in performing the contract which we would not otherwise have incurred and which is beyond that otherwise provided for in or reasonably contemplated by the contract. We expect to be entitled to an increase in the contract sum under clause 43.

Yours faithfully

5 Payment

Most contractors would say that securing payment is one of the trickiest parts of their job. All the standard forms dealt with in this book, with the exception of GC/Works/1 (1998), make provision for the contractor to terminate its employment if the employer does not honour certificates (or in the case of DB pay the amount in the application) within the due time. If the architect simply does not issue a certificate at the proper time, it is a breach of contract for which the employer may be liable. It may be possible to recover payment due without a certificate in such instances.

More difficult is the situation in which the architect certifies less than the contractor considers is due. If persuasion fails, the remedy is to refer the dispute to adjudication or arbitration. There is a difficulty in that arbitration can be a lengthy process and the architect can rectify an under-certification when issuing subsequent certificates before the arbitration takes place. Even the relatively speedy adjudication will usually take 35 days from start to finish, which is quite long enough for the architect to issue another certificate. However, that should not preclude the contractor from recovering damages for the late certification. That will usually include interest.

The following letters deal with several common situations, including failure of the architect to issue the final certificate on the due date. Special letters are included to cover the special payment situations under DB.

Letter 131
To architect, enclosing interim application for payment (a)
This letter is only suitable for use with DB.

Dear

Under the provisions of clause 4.9, we enclose an application for interim payment stating the amount due to us calculated in accordance with clause 4.13 [*if alternative A applies or substitute '4.14' if alternative B applies*]. In support of our application we enclose the following details as required under clause 4.9.3:

[*List*]

We draw your attention to clause 4.10.1 which stipulates that you must pay the amount stated as due in this application within 14 days of its receipt.

Yours faithfully

Letter 132
To architect, enclosing interim application for payment (b)
This letter is only suitable for use with MW, MWD or
GC/Works/1 (1998).

Dear

We enclose an application for progress payment [*substitute 'payment of advance on account' when using GC/Works/1 (1998)*] stating the amount due to us calculated in accordance with the provisions of clause 4.3 [*substitute '48' when using GC/Works/1 (1998)*]. In support of our application, we enclose the following documents:

[*List*]

Yours faithfully

Letter 133
To quantity surveyor, submitting valuation application
This letter is only suitable for use with SBC, IC or ICD.

Dear

We enclose an application setting out what we consider to be the valuation calculated in accordance with clause 4.10 [*substitute '4.7' when using IC or ICD*].

May we remind you that, under clause 4.12 [*substitute '4.6.3' when using IC or ICD*], you must now make a valuation for the purpose of ascertaining the amount due in an interim certificate and, to the extent that you disagree with our application, you must submit to us a statement which identifies such disagreement. Note that the statement must be in similar detail to that contained in our application.

Yours faithfully

Letter 134
To architect, if quantity surveyor fails to respond to the valuation application
This letter is only suitable for use with SBC, IC or ICD.

Dear

We acknowledge receipt of certificate number [*insert number*] dated [*insert date*] from which we see that the amount valued was substantially less than the amount included in our application submitted under the provisions of clause 4.12 [*substitute '4.6.3' when using IC or ICD*]. We are concerned because, contrary to the express provisions of the clause, the quantity surveyor has not submitted to us a statement, in similar detail to our application, identifying any disagreement with our application. This should have been done at the time of the quantity surveyor's valuation, i.e., no later than 7 days before the date of the next certificate.

The quantity surveyor's failure amounts to a breach of contract for which we are entitled to appropriate damages. When we did not receive the quantity surveyor's detailed disagreement by the valuation date, we were entitled to assume that the quantity surveyor did not disagree. Accordingly and in reliance, we undertook expenditure of sums of money which we now realise we are not going to receive.

We suggest that damages could be avoided if you would immediately issue a supplementary certificate for the difference between the two valuations.

Yours faithfully

Letter 135
To architect, if interim certificate not issued
This letter is not suitable for use with DB.

Dear

We have not received a copy of the interim certificate which should have been issued on the [*insert date*] in accordance with clause 4.9.1 [*substitute '4.6.1' when using IC or ICD, '4.3' when using MW or MWD or '50(1)' when using GC/Works/1 (1998)*] of the conditions of contract.

It may be that the certificate has been lost in the post and on this assumption we should be pleased if you would send us a further copy.

If you have not, in fact, issued a certificate, we must remind you of your contractual duty so to do and request that it is in our hands by [*insert date*]. Failing which, we will take immediate legal action against the employer for the breach.

Yours faithfully

Copy: Employer

Letter 136
To architect, if certificate insufficient
This letter is not suitable for use with DB.

Dear

We have received your interim certificate number [*insert number*] dated [*insert date*]. We note [*insert the disputed figures*] which do not correspond with the evidence in the documents we have submitted, our discussions with you/the quantity surveyor [*delete as appropriate*] or the situation on site.

If we do not receive a corrected certificate by [*insert date*] we shall seek immediate adjudication on the matter.

Yours faithfully

Copy: Employer

Letter 137
To employer, if payment not made in full and no withholding notice issued
Special/recorded delivery

Dear

We have today received your cheque for [*insert amount*], some [*insert difference*] less than certified due to us in the architect's certificate number [*insert number*] dated [*insert date*]. We note that you have withheld [*insert amount*], but you have given no reasons for doing so/the notice you have sent is out of time/the notice you have sent does not properly particularise the grounds for withholding [*delete as appropriate*].

Therefore, your action amounts to a breach of contract and it is contrary to the provisions of the Housing Grants, Construction and Regeneration Act 1996. If we do not receive the sum of [*insert the amount withheld*] by [*insert date*] we will exercise our contractual or other remedies as we deem appropriate to recover the full amount due.

Yours faithfully

Copy: Architect

Letter 138
To architect, regarding copyright if payment withheld
This letter is only suitable for use with SBC, ICD, MWD or DB.

Dear

We note that you persist in withholding payment of [*describe the situation*].

[*When using MWD add:*]

Copyright in the design of the contractor's designed portion vests in this company.

[*Otherwise add:*]

Clause 2.41 [*substitute '23.3' when using ICD or '2.38' when using DB*] expressly provides that copyright in all designs, drawings and other documents which we provide vests in this company.

[*Then add:*]

You will only acquire a licence to reproduce such designs in the form of a building when you have paid the amounts you owe. Until then and for the avoidance of doubt take this as notice that you are currently infringing our copyright and we do not grant you a licence. If you do not pay us by close of business on [*insert date*] the money owing, we shall seek to recover substantial damages from you in respect of the infringement.

Yours faithfully

Letter 139
To employer, if the advance payment is not paid on the due date
This letter is not suitable for use with MW, MWD or GC/Works/1
(1998).

Dear

Under the provisions of clause 4.8 [*substitute '4.5' when using IC or ICD or '4.6'*
when using DB] an advance payment in the sum of [*insert the amount*] was due
to be paid to us on the [*insert date*] as stated in the contract particulars. That is
[*insert number*] days ago. We furnished a bond on the [*insert date*] in the
standard terms from a surety which you have approved.

Your failure to provide the advance payment is a serious breach of contract.
We were relying on the payment to assist our funding of this project. Indeed,
the offer of such payment was the deciding factor in deciding to enter into this
contract.

We are advised that we are entitled to damages for your breach and that, if it
continues for more than a few days, we may be able to treat it as repudiatory,
because it will effectively prevent us from proceeding with the Works. In that
case, we would be able to accept the breach and bring our obligations to an
end. Hopefully, that will not be necessary and we look forward to receiving
your payment by return.

Yours faithfully

Letter 140
To architect, if valuation not carried out in accordance with the priced activity schedule
This letter is only suitable for use with SBC, IC or ICD.

Dear

We have just received certificate number [*insert number*] dated [*insert date*] from which we see that the amount certified is substantially less than we expected. It appears that the valuation has not been carried out using the priced activity schedule we supplied.

May we draw your attention to clause 4.16.1.1 [*substitute '4.7.1.1' when using IC or IDC*] which specifically provides that where there is an activity schedule, the value of work in each activity to which it relates must be a proportion of the price stated for the work in that activity equal to the proportion of the work in that activity that has been properly executed.

A simple breakdown is enclosed showing the figure produced by applying that calculation. No doubt the matter is simply the result of an oversight. However, in view of the serious shortfall in the amount certified, we believe that a further and immediate supplementary certificate should be issued to rectify the position.

Yours faithfully

Letter 141
To architect, if contractor not asked to be present at measurement
This letter is only suitable for use with SBC or GC/Works/1 (1998).

Dear

Looking at the latest valuation, we are concerned to note that although it obviously concerns work which the quantity surveyor has had to measure for valuation purposes, we were not given the opportunity to be present while the measurement was carried out. This is contrary to [*insert 'intention of' when using GC/Works/1 (1998)*] the provisions of clause 5.4 [*substitute '18(1)' when using GC/Works/1 (1998)*]. The work to which we refer is [*name or describe the work*].

The work in question is now covered up. We should be pleased to receive copies of your measurement notes. To the extent that we do not agree your measurements, we believe that your breach leads to a presumption that our quantities are correct.

We look forward to hearing from you.

Yours faithfully

Copy: Quantity surveyor

Letter 142
To architect, requesting payment for off-site materials
This letter is only suitable for use with SBC, IC or ICD.

Dear

The following goods and materials are 'Listed Items' and are stored at [*insert place*] and available for your inspection at any time:

[*List*]

We should be pleased if you would operate the provisions of clause 4.17 [*substitute '4.12' when using IC or ICD*] of the conditions of contract and include the value of such goods and materials in your next interim valuation. We confirm that we have complied with all the requirements of the contract in respect of such goods and materials which the enclosed documents prove are our property.

[*Add, if appropriate:*]

We confirm that we have provided a bond in terms as annexed to the contract.

Yours faithfully

Letter 143
To employer, giving 7 days notice of suspension
Special delivery

Dear

We note that you have not issued any withholding notice, but you have failed to pay/pay in full [*delete as appropriate*] the sum due by the final date for payment. Without prejudice to our other rights and remedies, we give notice that unless you pay the amount due in full within 7 days after receipt of this letter, we shall, in accordance with clause 4.14 [*substitute '4.11' when using IC, ICD or DB, '4.7' when using MW or MWD or '52' when using GC/Works/1 (1998)*], suspend performance of all our obligations under the contract until full payment is received.

Yours faithfully

Copy: Architect [*only when using SBC, IC, ICD, MW or MWD*]

Letter 144
To employer, if payment in full has not been made within 7 days despite notice of suspension

Dear

Further to our letter dated [*insert date*] stating that if payment in full was not made within 7 days from the date of its receipt we would suspend our obligations, we have not received any/full [*delete as appropriate*] payment. This is to inform you that, with immediate effect, we are suspending all our obligations under the contract.

We have left the site safe, but it is now for you to arrange security and insurance if you persist in withholding payment.

Yours faithfully

Letter 145
To employer, requesting interest on late payment
This letter is not suitable for use with GC/Works/1 (1998).

Dear

We refer to certificate number [*insert number*] dated [*insert date*]. The final date for payment was [*insert date*]. At the time of writing we have not received any/full [*delete as appropriate*] payment.

Under the provisions of clause 4.13.6 [*substitute '4.8.5' when using IC or ICD, '4.4' when using MW or MWD and '4.10.6' when using DB*] you are obliged to pay us simple interest at 5% above Bank of England Base Rate current at the date payment became overdue until the amount is paid. However, you should note that this contractual right to interest is without prejudice to our other rights and remedies. This letter is simply by way of notice that we have no intention of waiving our right to interest although we intend to write to you under separate cover if payment is not made by close of business on the day following the date of this letter.

Yours faithfully

Letter 146

To employer, requesting retention money to be placed in a separate bank account
This letter is not suitable for use with MW, MWD or GC/Works/1 (1998).
Special delivery

Dear

We formally request you to place all current and future retention money in a separate bank account set up for the express purpose and identified as money held in trust for our benefit. Please inform us of the name of the bank, the account name and number. Notwithstanding the provisions of the contract, it is established that your obligation exists irrespective of any formal request we may make.

Yours faithfully

Copy: Architect

Letter 147

To employer, if failure to place retention in separate bank account
This letter is not suitable for use with MW, MWD or GC/Works/1 (1998).
Special delivery

Dear

Further to our letter dated [*insert date*] you have not notified us that you have set aside retention money in a separate trust fund as we requested and as the contract and the law provides.

The separate trust fund will protect our money in the event of your insolvency. If you have not complied with our request by [*insert date*] we shall immediately seek an injunction to compel you to comply.

Yours faithfully

Copy: Architect

Letter 148

To architect, enclosing all information for preparation of final certificate
This letter is not suitable for use with DB or GC/Works/1 (1998).

Dear

In accordance with clause 4.5.1 [*substitute '4.13.1' when using IC or ICD or '4.8.1' when using MW or MWD*] we enclose full details of the final account for this contract together with all supporting documentation.

We should be pleased if you would proceed with the necessary calculations and verifications to enable the final certificate to be issued in accordance with the contract timescale.

Yours faithfully

Letter 149
To architect, enclosing final account
This letter is only suitable for use with DB.

Dear

In accordance with clause 4.12.1 of the conditions of contract we enclose the final account and final statement referred to in clause 4.12.3 and we should be pleased to have your agreement.

We enclose the following supporting documentation:

[*List*]

If you reasonably require any further information, we will be pleased to provide it. Please let us have a complete list of such requirements (if any) within one month from the date of this letter.

Yours faithfully

Letter 150
To employer, disputing the final account
This letter is only suitable for use with DB.

Dear

We are in receipt of the employer's final account and employer's final statement.

Take this as formal notice, under the provisions of clause 4.12.7, that we dispute the entirety of such employer's final account and employer's final statement. Therefore, the balance will not become conclusive.

We suggest that a meeting is appropriate to agree the final account and we would be free to attend such a meeting on [*insert dates*] at [*insert times*].

Yours faithfully

Letter 151
To architect, if final certificate not issued on the due date (a)
This letter is only suitable for use with SBC.
Special delivery

Dear

Clause 30.8 of the conditions of contract requires you to issue the final certificate within 2 months from the latest of the following events:

1. The end of the rectification period/last rectification period [*delete as appropriate*] – [*insert date*].
2. Date of issue of the certificate of making good/last certificate of making good [*delete as appropriate*] – [*insert date on certificate*].
3. The date on which you sent us a copy of the statement and ascertainment under clause 4.5.2 – [*insert date or, if the architect has not sent an ascertainment and statement, add the following:*] In this instance, you have not sent us such ascertainment and statement which should have been issued on the [*insert date*].

Therefore, the final certificate should have been issued on the [*insert date calculated as above*]. Some [*insert number*] weeks have passed since that date and we have received no such certificate. You are in breach of contract, a breach for which, we are advised, the employer will become liable after receipt of this notification. Without prejudice to our rights in this matter, if the final certificate is in our hands by [*insert date*], we will take no further action on such breach.

Yours faithfully

Copy: Employer

Letter 152

To architect, if final certificate not issued on the due date (b)
This letter is only suitable for use with IC or ICD.
Special delivery

Dear

Clause 4.14.1 of the conditions of contract requires you to issue the final certificate within 28 days of the latest of the following events:

1. The sending to us of the computations of the adjusted contract sum, which we received on the [*insert date*].
2. Your certificate of making good under clause 2.31 – [*insert date*].

Therefore, the final certificate should have been issued on the [*insert date calculated as above*]. Some [*insert number*] weeks have passed since that date and we have received no such certificate. You are in breach of contract, a breach for which, we are advised, the employer will become liable after receipt of this notification. Without prejudice to our rights in this matter, if the final certificate is in our hands by [*insert date*], we will take no further action on such breach.

Yours faithfully

Copy: Employer

Letter 153

To architect, if final certificate not issued on the due date (c)
This letter is only suitable for use with MW or MWD.
Special delivery

Dear

Clause 4.8.1 of the conditions of contract requires you to issue the final certificate within 28 days of receipt by you of all documentation reasonably required for computation of the amount to be certified, provided that you have issued your certificate under clause 2.11 [*substitute '2.12' when using MWD*].

You confirmed that all such documents were in your possession by your letter of the [*insert date*]. You issued a clause 2.11 [*substitute '2.12' when using MWD*] certificate on the [*insert date*].

Therefore, the final certificate should have been issued on the [*insert date calculated as above*]. Some [*insert number*] weeks have passed since that date and we have received no such certificate. You are in breach of contract, a breach for which, we are advised, the employer will become liable after receipt of this notification. Without prejudice to our rights in this matter, if the final certificate is in our hands by [*insert date*], we will take no further action on such breach.

Yours faithfully

Copy: Employer

6　Extensions of Time

The whole business of delays and extensions of time is widely misunderstood. The principal reason for the existence of an extension of time clause is to enable a known completion date to be fixed in response to any delay for which the employer either is responsible or has taken responsibility so that liquidated damages can be recovered thereafter. A purely subsidiary reason is to take some of the risk from the contractor for matters which are outside the control of both parties. All the standard forms, with the exception of MW and MWD, make detailed provision for what is to happen if delay occurs. In order to ensure that you obtain the extensions to which you are entitled under a particular contract, you must be scrupulous in complying with the detailed provisions for notices and supporting information. Although your notice of delay is not a condition precedent to the granting of an extension of time, failure to observe the provisions properly often results in extensions being granted later rather than sooner and for rather less a period than expected.

The following letters deal with notification of delay and some of the problems which commonly arise, often due to the architect or the contractor being imperfectly aware of their obligations.

Letter 154
To architect, if delay occurs, but no grounds for extension of time
This letter is not suitable for use with MW, MWD or GC/Works/1 (1998).

Dear

The progress of the Works is being/is likely to be [*delete as appropriate*] delayed due to [*state reasons*].

We will continue to use our best endeavours to minimise the delay and its effects and we will inform you immediately the cause of the delay has ceased to operate.

This notice is issued in accordance with clause 2.27.1 [*substitute '2.19.1' when using IC or ICD or '2.24.1' when using DB*].

Yours faithfully

Letter 155

To architect, when cause of delay ended if no grounds for extension of time
This letter is not suitable for use with MW, MWD or GC/Works/1 (1998).

Dear

We refer to our letter of the [*insert date*] notifying you of delay.

We are pleased to be able to inform you that the cause of the delay has been dealt with. The measures which we adopted during the period of delay and the continuing procedures over the next few weeks are designed to enable us to recover the lost time and put the progress of the Works on programme by [*insert date*].

Yours faithfully

Letter 156
To architect, if delay occurs giving grounds for extension of time (a)
This letter is not suitable for use with MW, MWD or GC/Works/1 (1998).

Dear

The progress of the Works is being delayed due to [*state reasons*]. We consider this to be a relevant event under clause [*insert number*].

The delay began on [*insert date*]. When it is finished, we will furnish you with our estimate of delay in the completion of the Works and further supporting particulars.

You may be assured that we are using our best endeavours to prevent delay in progress and completion of the Works.

This notice is issued in accordance with clause 2.27.1 [*substitute '2.19.1' when using IC or ICD or '2.24.1' when using DB*].

Yours faithfully

Letter 157
To architect, if delay occurs giving grounds for extension of time (b)
This letter is only suitable for use with MW, MWD or GC/Works/1 (1998).

Dear

It is apparent that the Works will not be complete by the date for completion.

[*Then either:*]

The circumstances are [*state*]. We estimate the delay to total [*insert period*]. We consider that we are entitled to an extension of [*insert period*] and we will furnish further particulars within the next few days.

[*Or:*]

The circumstances are [*state*]. We are unable to estimate the total delay at this stage because it is continuing. When the delay is over, we will submit the required estimate and other details immediately.

[*Then:*]

You may be assured that we are using our best endeavours to prevent delay in progress and completion of the Works.

This notice is issued in accordance with clause 2.7 [*substitute '2.8' when using MWD or '36(1)' when using GC/Works/1*] of the conditions of contract.

Yours faithfully

Letter 158

To architect, providing further particulars for extension of time (a)
This letter is not suitable for use with MW, MWD or GC/Works/1 (1998).

Dear

We refer to our letter of the [*insert date*] in which we notified you of a delay in progress of the Works likely to result in a delay to completion of the Works. We note below particulars of the expected effects and the estimated extent of delay in completion of the Works in respect of each relevant event specified in our notice:

[*List relevant events separately, giving an assessment of the delay to completion in each case together with any other supporting information.*]

We believe that you now have sufficient information to enable you to grant a fair and reasonable extension of time.

Yours faithfully

Letter 159
To architect, providing further particulars for extension of time (b)
This letter is only suitable for use with MW, MWD or GC/Works/1 (1998).

Dear

We refer to our notice of the [*insert date*] in which we informed you of a delay in progress of the Works likely to result in a delay to completion of the Works.

We estimate that completion will be delayed by [*insert number*] weeks and we consider that we should be granted an extension of time for that period. We arrive at this conclusion as follows:

[*State reasons in full and include all supporting information.*]

We believe that you now have sufficient information to enable you to grant a fair and reasonable extension of time and we look forward to hearing from you shortly.

Yours faithfully

Letter 160
**To architect, if requesting further information in order to grant
extension of time**

Dear

Thank you for your letter of the [*insert date*] in which you request further
information in respect of [*state what architect requires*].

[*State the information required by the architect in the form of precise answers to the
questions or, if the architect does not say what is required, write as follows:*]

We believe that we gave you all the information you require in our letter of the
[*insert date*]. It is not clear from your letter what further information you now
request. If you would be good enough to ask specific questions, we will do our
best to answer them and supply whatever further supporting details then
suggest themselves to us.

We look forward to hearing from you as soon as possible.

Yours faithfully

Letter 161
To architect, if unreasonably requesting further information in order to grant an extension of time

Dear

Thank you for your letter of the [*insert date*] requesting further information in order to enable you to grant an extension of time.

We submitted notice of delay [*substitute 'request for extension of time' when using GC/Works/1 (1998)*], as required by the contract, on [*insert date*]. We submitted full particulars including estimate of the effect of the delay on completion date on [*insert date or dates or substitute 'at the same time' if appropriate*]. We believe that you had all the information necessary to enable you to make a fair and reasonable extension of time by [*insert date*]. It is, of course, very much in our interests to supply you with full information as quickly as possible; this we have done.

It is our view that your latest request for information is nothing but an attempt to postpone the granting of an extension. We, therefore, formally call upon you to carry out your duty under clause 2.28.1 [*substitute '2.19.1' when using IC or ICD, '2.7' when using MW, '2.8' when using MWD or '36(1)' when using GC/Works/1 (1998)*].

Yours faithfully

Letter 162
To architect, if extension of time is insufficient

Dear

We have received today your notification of an extension of time of [*insert period*] producing a new date for completion of [*insert date*].

We find your conclusions inexplicable in the light of the facts and the information we submitted in support of those facts.

Perhaps you would be good enough to reconsider your grant of extension of time or let us have an indication of your reasons for arriving at the time period you have granted.

Yours faithfully

Letter 163
To architect, who is not willing to reconsider an insufficient extension of time

Dear

Thank you for your letter of the [*insert date*] from which we note that you are not willing to reconsider your grant of extension of time in response to our notice of the [*insert date*] and submissions of further information of the [*insert date*] in response to your request of the [*insert date*].

[*If the architect has given reasons:*]

We have carefully examined the reasons you give in support of your decision and they reveal that you have ignored much of our submission and the facts of the matter.

[*If the architect has not given reasons:*]

We note that you refuse to give any reasons for your decision and we can only assume that you are on uncertain ground.

[*Then add:*]

We will be happy to meet you if you think that a full discussion would be helpful. That appears to be the sensible way forward, but failing that, we intend to refer this dispute to immediate adjudication.

Yours faithfully

Letter 164
To architect, if extension of time not properly attributed
This letter is only suitable for use with SBC or DB.

Dear

Thank you for your notification of revision to the completion date. Unfortunately, you failed to state the extension of time you have attributed to each relevant event. This is now mandatory under clause 2.28.3 [*substitute '2.25.3' when using DB*].

Obviously, having formed your opinion, these details will be readily to hand and we look forward to receiving a revised notification containing this information within the next couple of days.

Yours faithfully

Letter 165
To architect, if extension of time not granted within time stipulated
This letter is not suitable for use with IC, ICD, MW or MWD.

Dear

Notice of delay [*substitute 'Request for extension of time' when using GC/Works/1 (1998)*] was sent to you on [*insert date*] in accordance with clause 2.27.1 [*substitute '2.24.1' when using DB or '36(1)' when using GC/Works/1 (1998)*] of the conditions of contract. Full particulars including estimate of the delay to completion and estimate of the extension required were sent to you on [*insert date*]. You made no request for further information. Clause 2.28.2 [*substitute '2.25.2' when using DB or '36(1)' when using GC/Works/1 (1998)*] requires you to notify us of your decision not later than 12 weeks [*substitute '42 days' when using GC/Works/1 (1998)*] from receipt of particulars [*substitute 'notice' when using GC/Works/1 (1998)*]. The period elapsed on [*insert date*] and you have not informed us of your decision. You are clearly in breach of contract, a breach for which the employer is responsible.

You are not empowered to make any extension of time for the current relevant events [*substitute 'circumstances' when using GC/Works/1 (1998)*] until after the contract completion date [*substitute 'completion of the Works' when using GC/Works/1 (1998)*]. Any loss or expense which we may suffer, whether from increasing resources or otherwise, as a result of your breach will be recovered from the employer as damages in due course.

Yours faithfully

Copy: Employer

Letter 166
To architect, if slow in granting extension of time
This letter is only suitable for use with IC, ICD, MW or MWD.

Dear

Notice of delay was sent to you on [*insert date*] in accordance with clause 2.19.1 [*substitute '2.7' when using MW or MWD*] of the conditions of contract. Full particulars were sent to you on [*insert date*].

You have now had [*insert number*] weeks in which to make your decision and we now call upon you to grant us the extension of time to which we are entitled. If we do not receive your notice granting such extension by [*insert date*] you will be in breach of contract, a breach for which the employer will be responsible.

[*If using IC or ICD, add:*]

You will not be empowered to make any extension of time for the current events until after the date of practical completion. Any loss or expense which we may suffer, whether from increasing resources or otherwise, as a result of your breach will be recovered from the employer as damages in due course.

[*If using MW 98, add:*]

We will consider that time is 'at large' and the employer will have lost the right to deduct liquidated damages because there will be no date for completion from which liquidated damages can run and you will have lost your power to fix such a date. Our obligation will then be to finish the Works within a reasonable time.

Yours faithfully

Copy: Employer

Letter 167
To architect, if review of extensions not carried out (a)
This letter is only suitable for use with SBC or DB.

Dear

Clause 2.28.5 [*substitute '2.25.5' when using DB*] requires you to either:

1.   Fix a completion date later than that previously fixed, or
2.   Fix a completion date earlier than that previously fixed, or
3.   Confirm the completion date previously fixed.

You must carry out this duty, at latest, within 12 weeks after the date of practical completion. We are advised that, despite speculation to the contrary, the time period is mandatory. That period expired on [*insert date*]. You are in breach of your duty and our obligation now is simply to complete within a reasonable time. The employer has lost the right to deduct liquidated damages, because there is no date fixed for completion from which such damages can run and you have lost your power to fix such a date. Any attempt to deduct such damages will result in immediate legal action on our part.

Yours faithfully

Copy: Employer

Letter 168
To architect, if review of extensions not carried out (b)
This letter is only suitable for use with IC or ICD.

Dear

Clause 2.19.3 permits you to extend time whether upon reviewing previous decisions or otherwise and whether or not we have given notice of delay. This is a valuable power for you to review the situation after the works are finished.

However, you must carry out this duty within 12 weeks after the date of practical completion. We are advised that, despite speculation to the contrary, the time period is mandatory. That period expired on [*insert date*] and our obligation now is simply to complete within a reasonable time. The employer has lost the right to deduct liquidated damages, because there is no date fixed from which such damages can run and you have lost your power to fix such a date. Any attempt to deduct such damages will result in immediate legal action on our part.

Yours faithfully

Copy: Employer

Letter 169
To architect, if no final decision on extensions of time (c)
This letter is only suitable for use with GC/Works/1 (1998).

Dear

Clause 36(4) of the conditions of contract stipulates that you must come to a final decision on all outstanding and interim extensions of time within 42 days after completion of the Works. We are advised that, despite speculation to the contrary, the time period is mandatory. The period expired on [*insert date*] and you have not made a final decision on the request(s) for extension of time originally notified to you on [*insert date or dates*] and our obligation now is simply to complete within a reasonable time. The employer has lost the right to deduct liquidated damages, because there is no date fixed for completion and you have lost your power to fix such a date. Any attempt to deduct such damages will result in immediate legal action on our part.

Yours faithfully

Copy: Employer

Letter 170
To architect, who alleges that contractor is not using best endeavours
This letter is not suitable for use with MW or MWD.

Dear

Clause 2.28.6.1 [*substitute '2.19.4.1' when using IC or ICD, '2.25.6' when using DB or '36(6)' when using GC/Works/1 (1998)*] of the conditions of contract requires us to use our best endeavours to prevent delay. Your allegation in [*state where and date, e.g.: minute no. 7.4 of the site meeting held on the 3 September 2007*] that we are failing to carry out our duties in this respect is totally without foundation. Our obligation to use best endeavours is simply an obligation to continue to work regularly and diligently, rearranging our labour force as best we can. This we have done and we are continuing so to do. There is no obligation upon us to expend additional sums of money to make up lost time. If that was the case, the extension of time clause would be otiose.

If you purport to reduce, on grounds of failure to use best endeavours, our entitlement to an extension of time, we will take immediate and appropriate advice on the remedies available to us.

Yours faithfully

Letter 171

To architect, if non-completion certificate or notice wrongly issued
This letter is not suitable for use with MW, MWD or GC/Works/1 (1998)

Dear

We are in receipt of what purports to be a non-completion certificate [*substitute 'notice' when using DB*] issued under clause 2.31 [*substitute '2.22' when using IC or ICD or '2.28' when using DB*]. The issue of a valid certificate [*substitute 'notice' when using DB*] is a pre-condition to the deduction of liquidated damages. In this instance, it is not valid because

[*Either:*]

you have not yet given the correct extension of time.

[*Or:*]

it was issued prior to the completion date.

[*Or:*]

it does not comply with the terms of the contract.

[*Then add:*]

If the employer attempts to withhold liquidated damages, a dispute will then arise which we will refer to adjudication. In order to avoid the resultant expense to both parties, we suggest that you withdraw your certificate [*substitute 'notice' when using DB*] immediately.

Yours faithfully

Letter 172
To employer, if liquidated damages wrongfully deducted

Dear

We have received your cheque today in the sum of [*insert amount*] which falls short of the amount certified as due to us in certificate number [*insert number*] by [*insert amount*]. We note from your accompanying letter of the [*insert date*] that the deficit represents your deduction of alleged liquidated damages in the sum of [*insert amount*] for [*insert number*] weeks.

We consider you to be in breach of contract because [*state reasons*]. If we do not receive your cheque for the full amount of [*insert amount*] by [*insert date*], we will take appropriate steps to recover not only the amount wrongfully deducted, but also damages, interest and costs. We reserve the right to suspend performance of all obligations under clause 4.14 [*substitute '4.11' when using IC, ICD or DB, '4.7' when using MW or MWD or '52' when using GC/Works/1 (1998)*] and/or to terminate our employment under clause 8.9.1.1 [*substitute '6.8.1.1' when using MW or MWD. Omit this sentence when using GC/Works/1 (1998)*].

Yours faithfully

Letter 173
To employer, if damages repaid without interest

Dear

We are in receipt of your letter of the [*insert date*] enclosing your
cheque for [*insert amount*] representing liquidated damages wrongfully
deducted/recovered [*delete as appropriate*]. You will recall that in our letter
dated [*insert date*] we gave you due notice that we should require damages for
your breach. We are prepared to accept interest charges as a realistic basis for
damages and the additional payment required on account of such interest is,
therefore, calculated on the basis of 5% above Bank of England Base Rate
current at the date payment became overdue [*insert amount together with the
calculation*].

We assume that this is a genuine oversight on your part and, while reserving
all our rights and remedies, we do not propose to take any action if we receive
the sum of [*insert amount of interest*] by [*insert date*].

Yours faithfully

Copy: Architect

7 Loss and/or Expense

If you intend to claim loss and/or expense, you must carefully follow the procedure laid down in the contract. Failure to do so may result in your claim being rejected. Contractors sometimes do not realise that the architect has a duty to the employer, not to resist claims, but to accept only those claims which can be shown to be valid.

Most important is to make your application in good time. Remember, however, that there is nothing in any standard form contract which makes the obtaining of an extension of time a condition precedent before payment of direct loss and/or expense can be made. This has been confirmed by the courts.

If you cannot satisfy the provisions of the contract, and even if you can, there is nothing to prevent you from pursuing your claim for damages at common law. The standard forms to which reference is made in this book do not give the architect power to deal with common law claims.

The following letters deal with the initial application and the provision of further information. Common difficulties arise in regard to the speed of ascertainment and, of course, the amount of ascertainment. Some suggested letters to cover these situations are included.

Of course, if the architect will not certify loss and/or expense which you perceive to be due, you can refer the matter to the very quick adjudication process which is dealt with later in this book.

Letter 174
To architect, applying for payment of loss and/or expense (a)
This letter is not suitable for use with MW, MWD or GC/Works/1 (1998).

Dear

We hereby make application under clause 4.23 [*substitute '4.17' when using IC or ICD or '4.19' when using DB*] of the conditions of contract as follows:

We have incurred/are likely to incur [*delete as appropriate*] direct loss and/or expense and financing charges in the execution of this contract for which we will not be reimbursed by a payment under any other provision in this contract, because the regular progress of the Works has been/is likely to be [*delete as appropriate*] materially affected by [*describe and unless referring to deferment of possession, add:*] being a relevant matter in clause [*insert clause number*].

Yours faithfully

Letter 175

To architect, applying for payment of loss and/or expense under the supplemental provisions (b)
This letter is only suitable for use with DB.
Special delivery

Dear

We are entitled to an amount in respect of loss and/or expense (including financing charges) to be added to the contract sum in accordance with clause 4.19 of the conditions of contract. Our application for payment under clause 4.9 is attached. Therefore in accordance with supplemental provision paragraph 5.2 in Schedule 2, we submit our estimate of such loss and/or expense, incurred in the period immediately preceding that for which such application is made, which we require to be added to the contract sum. We shall continue to submit estimates in accordance with paragraph 5.3 for so long as we continue to incur direct loss and/or expense.

May we remind you that within 21 days of receipt of this estimate, you must give us written notice either that you accept our estimate, or that you wish to negotiate and in default of agreement that you wish to refer the issue as a dispute to adjudication, or clause 4.19 will apply.

Yours faithfully

Letter 176
To architect, applying for payment of loss and/or expense (c)
This letter is only suitable for use with MW or MWD.

Dear

We hereby give notice that regular progress of the Works has been materially disrupted/prolonged [*delete as appropriate*] by matters for which the employer or you as the employer's agent are responsible. In our view, such matters exceed the situation contemplated by clause 3.6.3 of the conditions of contract. The matters are [*describe*].

Although the contract appears to have no machinery for dealing with claims of this nature, we are advised that we may bring an action at common law. We believe that the matters are capable of easy resolution by the employer or by you with proper authority. We should be content to proceed on this basis and we should be pleased to hear whether the employer is in agreement. If the employer is not prepared to deal with us on this basis or if we do not hear from you by [*insert date*] we shall formulate our claim for damages at common law.

Yours faithfully

Copy: Employer

Letter 177
To architect, applying for payment of expense (d)
This letter is only suitable for use with GC/Works/1 (1998).

Dear

We hereby give notice under clause 46(3)(a) of the conditions of contract as follows:

The regular progress of the Works has been/is likely to be [*delete as appropriate*] disrupted/prolonged [*delete as appropriate*] due to [*describe*]. In consequence of such disruption/prolongation [*delete as appropriate*] we have properly and directly incurred/we will properly and directly incur [*delete as appropriate*] expense in performing the contract which we would not otherwise have incurred and which is beyond that otherwise provided for in or reasonably to be contemplated by the contract. We are/we expect to be [*delete as appropriate*] entitled to an increase in the contract sum under clause 46(1).

Yours faithfully

Letter 178
To architect, giving further details of loss and/or expense (a)
This letter is not suitable for use with MW, MWD or GC/Works/1 (1998).

Dear

Thank you for your letter of the [*insert date*] requesting further information in support of our application for loss and/or expense which was submitted to you on [*insert date*].

We enclose a copy of the programme in precedence diagram/network analysis [*delete as appropriate*] form, which we have marked up to show the circumstances in some detail. It can be seen that [*describe the circumstances in some detail, giving dates and times, numbers of operatives and names of key members of staff involved*].

Also enclosed is the following copy correspondence, extracts from the site diary and site minutes:

[*List with dates. Do not include the following paragraph when using DB:*]

We should be pleased if you would inform us if there are any particular points on which you require more information before you are able to form an opinion as required by clause 4.23 [*substitute '4.17' when using IC or ICD*] of the conditions of contract.

Yours faithfully

Letter 179
To architect or quantity surveyor, enclosing details of loss and/or expense (b)
This letter is only suitable for use with SBC, IC or ICD.

Dear

Thank you for your letter of the [*insert date*] requesting details of loss and/or expense in respect of the matters notified in our letter of [*insert date*].

We enclose the details together with supporting documentation. We should be pleased if you would proceed with the ascertainment of loss and/or expense as required by clause 4.23 [*substitute '4.17' when using IC or ICD*]. Please inform us immediately if you require any further information.

Yours faithfully

Letter 180
To quantity surveyor, providing information for calculation of expense (c)
This letter is only suitable for use with GC/Works/1 (1998).

Dear

We have pleasure in enclosing the documents listed below. They contain full details of all expenses incurred and evidence that the expenses directly result from the occurrence of one of the events described in clause 46(1).

We look forward to your decision in accordance with clause 46(5) within 28 days of receipt of this letter.

[*List documents and information.*]

Yours faithfully

Letter 181
To architect, if ascertainment delayed (a)
This letter is only suitable for use with SBC.

Dear

We refer to our notice of the [*insert date*] submitted under the provisions of clause 4.23 of the conditions of contract and our letters of the [*insert dates*] enclosing the further information you required in order to form an opinion and carry out ascertainment of the amount of loss and/or expense.

[*Insert number*] weeks have elapsed since we last submitted such information to you and, during that period, neither you nor the quantity surveyor have requested further information or details. Clause 4.4 imposes a clear duty on you to certify sums ascertained in whole or in part in the next interim certificate following ascertainment. Clause 4.23 states that 'as soon as' you are of the opinion that the Works are affected, you will 'from time to time thereafter' ascertain the amount of loss and/or expense. A similar provision is contained in clause 4.25. As soon as you have formed your opinion, you must begin the process of ascertainment. As soon as any sum has been ascertained, you must include the amount in the next interim certificate. Please inform us by [*insert date*] the amount which you have ascertained and intend to include in the next certificate.

Yours faithfully

Letter 182
To architect, if ascertainment delayed (b)
This letter is only suitable for use with IC or ICD.

Dear

We refer to our notice of the [*insert date*] submitted under the provisions of clause 4.17 of the conditions of contract and our letters of the [*insert dates*] enclosing the further information you required to form an opinion and carry out ascertainment of the amount of loss and/or expense.

[*Insert number*] weeks have elapsed since we last submitted such information to you and, during that period, neither you nor the quantity surveyor have requested further information or details. Clause 4.7.2 imposes a duty on you to certify ascertainment under clause 4.11 in your certification of interim payments 'to the extent that it has been ascertained'. Please inform us by [*insert date*] the amount which you have ascertained and intend to include in the next certificate.

Yours faithfully

Letter 183
To architect, if ascertainment too small
This letter is not suitable for use with MW, MWD or DB.

Dear

We refer to our notice dated [*insert date*] in respect of loss and/or expense [*substitute 'expense' when using GC/Works/1 (1998)*]. Your [*substitute 'The quantity surveyor's' when using GC/Works/1 (1998)*] letter of the [*insert date*] notifying us of the amount ascertained appears to take little account of the very full supporting information submitted by us on [*insert date or dates*].

Unless we hear from you by [*insert date*] that you will amend your ascertainment to take account of the information we have supplied, a dispute will have arisen which we will refer to adjudication in due course.

Yours faithfully

Letter 184
To employer, regarding a common law claim

Dear

We draw your attention to [*describe the circumstances giving rise to the claim with dates*]. These circumstances entitle us to a claim for damages against you.

The contract makes no provision for such a claim in such circumstances and we intend to take immediate steps to recover under the contractual dispute resolution procedures.

Yours faithfully

Copy: Architect

Letter 185
To employer, regarding a common law claim

WITHOUT PREJUDICE

Dear

We refer to the claim we have against you at common law arising from the circumstances notified to you in our letter of the [*insert date*].

If you are prepared to meet us to discuss our claim with a view to reaching a reasonable settlement of the matter, we will take no immediate legal steps.

Please inform us by [*insert date*] if you agree to this suggestion and let us know a date when such a meeting could take place.

Yours faithfully

Copy: Architect

8 Termination, Arbitration, Adjudication and Completion

Termination is a serious procedure and it presents a number of dangers. Not least of these is the fact that if you do not follow the contractual termination procedure exactly, you may be held to be repudiating the contract unlawfully and be liable for damages to the employer. Notices must be sent in the manner, form and to the person prescribed. Standard letters are included to cover the situation, but before actually issuing the notice of termination, you would be prudent to seek expert advice.

Arbitration is also serious and costly. By the time arbitration is considered, you should be receiving proper advice on your legal and contractual position, but some standard letters for seeking concurrence in the appointment of an arbitrator are included.

Adjudication in construction contracts was introduced by the Housing Grants, Construction and Regeneration Act 1996 (the Construction Contracts (Northern Ireland) Order 1997 in Northern Ireland). Some useful letters are included, but it is impossible to cover all the many complex issues which may arise. Adjudication is being used for extremely complex disputes, with large amounts at risk, for which it was never intended. If you are likely to be involved in that kind of dispute or a dispute which will require legal input, be sure to obtain proper assistance. It is generally unlikely that the employer will initiate an adjudication against you, although it does occasionally happen. Usually, you will initiate it.

The remaining letters in this section deal with completion of the Works and the rectification period.

Letter 186
To employer or architect, if default notice served
This letter is not suitable for use with GC/Works/1 (1998).
Special delivery

Dear

We are in receipt of your letter of the [*insert date*], which apparently you intend to be a default notice in accordance with clause 8.4.1 [*substitute '6.4.1' when using MW or MWD*] of the conditions of contract.

[*Add either:*]

The purported notice contains serious error.

[*Or:*]

We are advised that your purported notice is ambiguous.

[*Or:*]

The substance of the default specified in your purported notice is incorrect.

[*Or:*]

The default specified in your purported notice is the result of your own default. [*Explain as appropriate.*]

[*Or:*]

Your purported notice is wrongly served.

[*continued*]

Letter 186 continued

[*Then add:*]

Your notice is, therefore, invalid and of no effect. Take this as notice that if you proceed to give notice of termination, it will be unlawful, it will amount to repudiation and we will take immediate proceedings against you/the employer [*delete as appropriate*].

Yours faithfully

Copy: Architect/employer [*delete as appropriate*]

Letter 187
To employer or architect, if default notice served justly
This letter is not suitable for use with GC/Works/1 (1998).
Special delivery

Dear

We are in receipt of your letter of the [*insert date*] which you sent as a default notice in accordance with clause 8.4.1 [*substitute '6.4.1' when using MW or MWD*] of the conditions of contract.

We regret that you have felt it necessary to send such a notice, but we are pleased to be able to inform you that [*insert whatever steps are being taken to remove the default*].

Yours faithfully

Copy: Architect/employer [*delete as appropriate*]

Letter 188
To employer, if premature termination notice issued
This letter is not suitable for use with GC/Works/1 (1998).
Special delivery

Dear

We are in receipt of your notice dated [*insert date*] purporting to terminate our employment under clause 8.4.2 [*substitute '6.4.2' when using MW or MWD*] of the conditions of contract.

Your original default notice was received on the [*insert date*]. The Post Office will be able to confirm to you the date of delivery. Your notice of termination was, therefore, premature and of no effect and may amount to a repudiation of the contract for which we can claim substantial damages. We have already corrected the default specified in your original notice and, without prejudice to any of our rights and remedies in this matter and particularly (but without limitation) our right to treat your purported termination as repudiation, we will continue to work normally while we take appropriate advice.

Yours faithfully

Letter 189
To employer who terminates after notification of cessation of terrorism cover
This letter is not suitable for use with MW, MWD or GC/Works/1 (1998)

Dear

We are in receipt of your termination notice given under clause 6.10.2.2. You should note that, although clause 6.10.2.3 appears to preclude further payment, it is now established that it does not remove your obligation to pay, by the final date for payment, any amount due.

Yours faithfully

Letter 190
To employer, giving notice of default before termination
This letter is not suitable for use with GC/Works/1 (1998).
Special delivery

Dear

We hereby give you notice under the provisions of clause 8.9.1/8.9.2 [*delete as appropriate or substitute '6.8.1/6.8.2' when using MW or MWD*] of the conditions of contract that you are in default/a specified suspension event has occurred [*delete as appropriate*] in the following respect:

[*Insert details of the default with dates if appropriate and refer to the appropriate sub-clause.*]

which must be ended.

If you continue the default/specified suspension event [*delete as appropriate*] for 14 [*substitute '7' when using MW or MWD*] days after receipt of this notice, we may forthwith terminate our employment under this contract without further notice.

Yours faithfully

Letter 191
To employer, terminating employment after default notice
This letter is not suitable for use with GC/Works/1 (1998).
Special delivery

Dear

We refer to the default notice sent to you on the [*insert date*].

Take this as notice that, in accordance with clause 8.9.3 [*substitute '6.8.3' when using MW or MWD*], we hereby terminate our employment under this contract without prejudice to any other rights or remedies which we may possess.

We are making arrangements to remove all our temporary buildings, plant etc., and materials from the site and we will write to you again within the next week regarding financial matters.

Yours faithfully

Letter 192
To employer, terminating employment on the employer's insolvency
This letter is not suitable for use with GC/Works/1 (1998).
Special delivery

Dear

In accordance with the provisions of clause 8.10.1 [*substitute '6.9.1' when using MW or MWD*] of the conditions of contract, take this as notice that we hereby terminate our employment, because [*insert precise details of the insolvency event*]. This termination will take effect on receipt of this notice.

We are making arrangements to remove all our temporary buildings, plant etc., and materials from the site and we will write to you again within the next week regarding financial matters. This notice is without prejudice to any other rights and remedies we may possess.

Yours faithfully

Letter 193
To employer, where either party may terminate
This letter is not suitable for use with GC/Works/1 (1998).
Special delivery

Dear

The carrying out of the whole or substantially the whole of the uncompleted Works has been suspended for a continuous period of [*insert the relevant period from the contract particulars or insert 'one month' when using MW or MWD*] by reason of the following event(s): [*insert details of the event(s)*].

Take this as notice, in accordance with the provisions of clause 8.11.1 [*substitute '6.10.1' when using MW or MWD*] of the conditions of contract, that unless the suspension ceases within 7 days of receipt of this notice, we may terminate our employment under the contract.

Yours faithfully

Letter 194

To employer, terminating if suspension has not ceased after notice
This letter is not suitable for use with GC/Works/1 (1998).
Special delivery

Dear

Further to our notice served in accordance with the provisions of clause 8.11.1 [*substitute '6.10.1' when using MW or MWD*], the suspension has not ceased and we now hereby terminate our employment under the contract.

The consequences of such termination are set out in clause 8.12 [*substitute '6.11' when using MW or MWD*].

Yours faithfully

Letter 195
To employer, terminating employment after damage by insured risk
This letter is not suitable for use with MW, MWD or GC/Works/1 (1998).
Special delivery

Dear

We refer to our notice sent to you under the provisions of Schedule 3 [*substitute '1' when using IC or ICD*], paragraph C.4.1 on [*insert date*]. We consider that it is just and equitable to terminate our employment in accordance with paragraph C.4.4 and we hereby exercise our option to so terminate forthwith. [*The termination must be received by the employer not less than 28 days from the date of the occurrence of the loss or damage.*]

Yours faithfully

Letter 196
To employer, giving notice of intention to refer a dispute to adjudication
Special delivery

Dear

Under the provisions of clause 9.2 [*substitute '7.2' when using MW or MWD or '59(1)' when using GC/Works/1 (1998)*] and the Scheme for Construction Contracts (England and Wales) Regulations 1998 [*omit when using GC/Works/1 (1998)*] we intend to refer the following dispute or difference to adjudication: [*insert a description of the dispute including where and when it arose*]. We will be requesting the adjudicator to [*insert the nature of the redress sought e.g., 'order immediate payment of the outstanding amount of X or such sum as the adjudicator decides is due'*].

For the record, the names and addresses of the parties to the contract are as follows: [*set out the names and addresses which have been specified for the giving of notices*].

[*Either, if the adjudicator is named in the contract, add:*]

The adjudicator will be [*insert name*] as specified in the contract particulars/abstract of particulars [*delete as appropriate*].

[*Or, if the named adjudicator is unable to act or if none is named, add:*]

We are applying to [*insert the name of the nominating body*] for the nomination of an adjudicator.

Yours faithfully

Copy: Adjudicator/nominating body [*delete as appropriate*]

Letter 197
To nominating body, requesting nomination of an adjudicator
This letter is not suitable for use with GC/Works/1 (1998).
Special delivery

Dear

We enclose a notice of intention to refer a dispute and/or difference under the contract to adjudication. We have today served this notice and covering letter on the other party to the contract: [*insert the name of the employer*]. The contract was executed on [*insert the name of the contract as it appears on the cover*] terms. You are the selected nominator. Therefore, in accordance with clause 9.2 [*substitute '7.2' when using MW or MWD*] and paragraph 2 of the Scheme for Construction Contracts (England and Wales) Regulations 1998, we hereby make application to you to select a person to act as Adjudicator. A copy of the completed application form and the Referring Party's cheque in the sum of [*insert the amount*] is enclosed.

[*If appropriate, add:*]

We should be grateful if you would not nominate any of the following persons: [*list any adjudicators on the panel of the nominating body whom you do not wish to be nominated in this instance and give brief reasons*].

Yours faithfully

Copy: Employer

Letter 198
To adjudicator, enclosing the referral
Special delivery

Dear

We note that you are/have been nominated as [*delete as appropriate*] the adjudicator. In accordance with clause 9.2 [*substitute '7.2' when using MW or MWD or '59(2)' when using GC/Works/1 (1998)*] and paragraph 7 of the Scheme for Construction Contracts (England and Wales) Regulations 1998 [*omit when using GC/Works/1 (1998)*] we enclose our referral with this letter. Included are particulars of the dispute or difference, a summary of the contentions on which we rely, a statement of the relief or remedy sought and further material which we wish you to consider.

A copy of the referral and the accompanying documentation has been sent to [*insert name of employer and when using GC/Works/1 (1998) add the names of the PM and the QS*].

Yours faithfully

Copy: Employer [*add PM and QS when using GC/Works/1 (1998)*] with enclosures

[*The Referral must reach the adjudicator no later than 7 days after the date of the Notice of Adjudication.*]

Letter 199
To employer, if the adjudicator's decision is in your favour
Special delivery

Dear

We have today received a copy of the adjudicator's decision. We note that the decision is in our favour. You will be aware that you must comply with the adjudicator's decision <u>in accordance with the timescale laid down</u> [*if the adjudicator has not stated a time for compliance insert 'immediately on delivery of the decision in accordance with paragraph 21 of the Scheme' in place of the words underlined*].

[*If appropriate, add:*]

Therefore, please [*insert the adjudicator's decisions and relevant time scales converted to actual dates, e.g., 'pay us the sum of £40,000.00 by close of business on the 10 December 2007*].

If you fail to comply with the adjudicator's decision, we will immediately take enforcement proceedings through the courts, claiming interest, all our costs and expenses.

Yours faithfully

Letter 200
To employer, requesting concurrence in the appointment of an arbitrator
Special delivery

Dear

We hereby give you notice that we require the undermentioned dispute or difference between us to be referred to arbitration in accordance with article 8 [*substitute '7' when using MW or MWD or delete the reference to the article when using GC/Works/1 (1998)*] and clause 9.3 [*substitute '7.3' when using MW or MWD or 'clause 60' when using GC/Works/1 (1998)*] of the contract between us dated [*insert date*]. Please treat this as a request to concur in the appointment of an arbitrator.

The dispute or difference is [*insert brief description*].

I propose the following three persons for your consideration and require your concurrence in the appointment within 14 days of the date of service of this letter, failing which we shall apply to the President or Vice-President of [*insert the name of the appointor as set out in the contract particulars or the abstract of particulars as appropriate*].

[*List names and addresses of the three persons.*]

Yours faithfully

Letter 201
To appointing body, if there is no concurrence in the appointment of an arbitrator

Dear

We are contractors that have entered into a building contract on [*insert the name of the contract as it appears on the cover*] terms, clause 9.4.1 [*substitute 'Schedule 1, paragraph 1' when using MW or MWD or '60(1)' when using GC/Works/1 (1998)*] of which makes provision for your President or Vice-President to appoint an arbitrator in default of agreement.

We should be pleased to receive the appropriate form of application and supporting documentation, together with a note of the current fees payable on application.

Yours faithfully

Letter 202
To architect, if practical completion of the Works or a section imminent

Dear

We anticipate that the Works/section [*delete as appropriate*] will be complete on [*insert date*]. If you will let us know when you wish to carry out your inspection, we will arrange for M.. [*insert name*] to be on site to give immediate attention to any queries which may arise. We look forward to receiving your practical completion certificate [*substitute 'practical completion statement' when using DB or 'certificate that the Works/section are completed in accordance with the contract' when using GC/Works/1 (1998)*] following your inspection.

Yours faithfully

Letter 203
To architect, if completion certificate wrongly withheld (a)
This letter is not suitable for use with DB.

Dear

Thank you for your letter of the [*insert date*].

We are surprised to learn that, in your opinion, practical completion [*substitute 'completion' when using GC/Works/1 (1998)*] of the Works/section [*insert 'in accordance with the contract' when using GC/Works/1 (1998)*] has not been achieved. The items you list as outstanding can only be described as trivial. We do not consider that such items can possibly justify withholding your certificate.

We strongly urge you to reconsider the matter and to issue your certificate forthwith as required by clause 2.30 [*substitute '2.21' when using IC or ICD, '2.9' when using MW, '2.10' when using MWD or '39' when using GC/Works/1 (1998)*] of the conditions of contract. If we do not receive your certificate by [*insert date*] naming [*insert date*] as the date of practical completion [*substitute 'when the Works/section were completed in accordance with the contract' when using GC/Works/1 (1998)*], we will take whatever steps we deem appropriate to protect our interests.

The items you list are receiving attention in the normal way.

Yours faithfully

Letter 204
To architect, if completion statement wrongly withheld (b)
This letter is only suitable for use with DB.

Dear

Thank you for your letter of the [*insert date*].

We are surprised to learn that, in your opinion, practical completion of the Works/section [*delete as appropriate*] has not been achieved. The items you list as outstanding can only be described as trivial and they cannot possibly justify your contention.

We draw your attention to the fact that, under this form of contract, you have no certifying function and the issue of the practical completion statement is simply a process of recording a matter of fact. It is not something for your opinion. We strongly urge you to issue the written statement forthwith as required by clause 2.27 of the conditions of contract. If we do not receive your statement by [*insert date*] naming [*insert name*] as the date of practical completion, we will take whatever steps we deem appropriate to protect our interests.

The items you list are receiving attention in the normal way.

Yours faithfully

Letter 205
To employer, consenting to early use
This letter is not suitable for use with MW, MWD or GC/Works/1 (1998)

Dear

Further to your request dated [*insert date*] under clause 2.6.1 [*substitute '2.5.1' when using DB*] to use or occupy the Works before the date for completion, we have notified the insurers [*substitute 'you should notify the insurers' if the employer is insuring under option B or C*] to seek confirmation that such use or occupation will not prejudice the insurance.

[*If insurance options B or C apply, add:*]

Subject to such confirmation, we consent to such use or occupation as you describe.

[*If insurance option A applies, add:*]

If the insurers require an extra premium, our consent will be subject to your agreement under clause 2.6.2 [*substitute '2.5.2' when using DB*].

Yours faithfully

Letter 206
To employer, consenting to partial possession (a)
This letter is only suitable for use with SBC, IC or ICD.

Dear

In response to your letter of the [*insert date*], we consent to your request to take partial possession of the Works, namely [*describe part or parts*] provided:

1. The date for possession will be [*insert date*] and the architect will give a written statement to that effect on your behalf in accordance with clause 2.33 [*substitute '2.25' when using IC or ICD*].
2. [*Insert whatever particular conditions may be appropriate to the circumstances.*]

If you will, or you will authorise the architect to, write to us indicating agreement to the above conditions, we will make arrangements to hand over the appropriate keys on the [*insert date*].

Yours faithfully

Copy: Architect

Letter 207
To employer, consenting to partial possession (b)
This letter is only suitable for use with DB.

Dear

In response to your letter of the [*insert date*], we consent to your request to take partial possession of the Works, namely [*describe part or parts*] provided:

1. The date for possession will be [*insert date*].
2. [*Insert whatever particular conditions may be appropriate to the circumstances.*]

If you will write to us agreeing to the above conditions, we will make the necessary arrangements to hand over the appropriate keys on the [*insert date*]. We shall then issue a written statement in accordance with clause 2.30.

Yours faithfully

Letter 208
To employer, issuing written statement of partial possession
This letter is only suitable for use with DB.

Dear

Further to your letter of the [*insert date*] indicating agreement to the conditions contained in our letter of the [*insert date*], take this as the written statement which we are to issue in accordance with clause 2.30 of the conditions of contract. We identify the part(s) of the Works taken into possession (the relevant part(s)) as [*describe the part or parts in sufficient detail to allow no mistake*]. The date on which you took possession (the relevant date) was [*insert date*].

We draw your attention to the consequences of partial possession, particularly as they apply to the commencement of the rectification period, your insurance liability and the reduction in any liability which we may have for liquidated damages.

Yours faithfully

Letter 209
To employer, refusing consent to partial possession
This letter is not suitable for use with MW, MWD or GC/Works/1 (1998).

Dear

Thank you for your letter of the [*insert date*] requesting our consent to you taking partial possession of the Works, namely [*describe part or parts*].

We regret that we feel unable to give our consent in this instance, because [*insert reasons*].

[*If appropriate, add:*]

We will let you know immediately if circumstances change so substantially that we feel able to consent to your request.

Yours faithfully

Copy: Architect [*omit when using DB*]

Letter 210
To architect, after receipt of schedule of defects

Dear

Thank you for your instruction number [*insert number*] dated [*insert date*] scheduling the defects you require to be made good now that the rectification period [*substitute 'maintenance period' when using GC/Works/1 (1998)*] has ended.

We have carried out a preliminary inspection and we are making arrangements to make good most of the items on your schedule. However, we do not consider that the following items are our responsibility for the reasons stated:

[*List, giving reasons.*]

We shall, of course, be happy to attend to such items if you will let us have your written agreement to pay us daywork rates for the work.

Yours faithfully

Letter 211
To architect, when making good of defects completed

Dear

We are pleased to inform you that all making good of defects has been completed in accordance with your schedule [*if appropriate, substitute 'your amended schedule'*]. We should be pleased if you would carry out your own inspection and confirm your satisfaction [*when using SBC, IC, ICD, MW or MWD, add:*] by issuing a certificate of making good.

Yours faithfully

Letter 212
To architect, returning drawings, etc. after final payment if requested
This letter is not suitable for use with DB.

Dear

Thank you for your letter of [*insert date*].

We enclose, as requested, all copies of drawings, details, descriptive schedules and other documents of like nature which bear your name and which are in our possession. We have, naturally, retained our copy of the contract documents for record purposes.

Yours faithfully

9 Sub-Contractors

The final section deals with sub-contractors. Matters included are:

- Assignment
- Sub-letting
- Letter of intent
- Warranty
- Objections
- Inability to conclude a sub-contract
- Entering into the contract
- Insurance
- Drawings
- Directions
- Extension of time
- Loss and/or expense
- Payment
- Set-off and withholding notices
- Suspension and termination
- Practical completion
- Sub-contractor's design failure
- Sub-consultant professional indemnity, warranty and design matters.
- Adjudication

The provisions with regard to named persons under IC and ICD are complex and a number of the following letters relate to these contracts or to the associated sub-contract ICSub/NAM/C. To use the standard letters effectively, you must first have carefully read and digested the appropriate sub-contract and any related documentation.

Letter 213
To employer, requesting consent to assignment

Dear

In accordance with clause 7.1 [*substitute '3.1' when using MW or MWD or '61' when using GC/Works/1 (1998)*] we should be pleased to receive your consent to the assignment of our rights to payment under this contract to [*insert name*].

We wish to take such action, because [*give reasons briefly*]. We acknowledge that our obligations under the contract will be unaffected.

Yours faithfully

Letter 214
To employer, if asked to consent to assignment
This letter is not suitable for use with GC/Works/1 (1998).

Dear

Thank you for your letter of the [*insert date*] from which we understand that you wish to assign your rights/certain of your rights [*delete as appropriate*] under the contract.

We are anxious to assist you if we can and we may be able to give our consent, subject to certain safeguards. Clearly, this is not something on which we can make an immediate decision and we are taking advice. We expect to be in a position to write to you again very soon.

Yours faithfully

Letter 215
To sub-contractor, assessing competence under the CDM Regulations

Dear

In accordance with Regulation 4 of the CDM Regulations 2007, we are obliged to take steps to satisfy ourselves that you are competent. [*Where the contractor and sub-contractor have a long and on-going business relationship add:*] Although, in your case, this is little more than a formality we have to show that we have taken reasonable steps. [*Then continue:*]

The Approved Code of Practice: 'Managing health and safety in construction' states that competence should be assessed in relation to certain core criteria which have been agreed by industry and HSE. We attach a copy of these criteria which form Appendix 4 of the Code. We should be grateful if, by the [*insert date*], you would furnish the evidence which we have highlighted in the third column. The Code states that 'unnecessary bureaucracy associated with competency assessment obscures the real issues . . . ' and we have endeavoured to take this into account when requesting evidence.

Alternatively, if you are subject to an independent accreditation organisation which assesses competence against these criteria, please provide relevant details.

Yours faithfully

Letter 216
To a designer, assessing competence under the CDM Regulations
This letter is only suitable for use with SBC, ICD, MWD, DB and GC/Works/1 (1998).

Dear

In accordance with Regulation 4 of the CDM Regulations 2007, we are obliged to take steps to satisfy ourselves that you are competent. [*Where the contractor and sub-contractor have a long and on-going business relationship add:*] Although, in your case, this is little more than a formality we have to show that we have taken reasonable steps. [*Then continue:*]

The Approved Code of Practice: 'Managing health and safety in construction' gives guidelines for establishing competency.

Therefore, we should be grateful if you would provide details of your membership of professional bodies and institution and, if applicable, the grade of membership. Please also apply details of your past experience of designing work of a similar kind and value as this project. The names of referees to whom we could refer would be useful. Please let us have this information by the [*insert date*].

Yours faithfully

Letter 217
To architect, requesting consent to sub-letting

Dear

We propose to sub-let portions of the Works [*substitute 'design' if appropriate when using SBC, ICD, MWD or DB*] as indicated below, because [*state reasons*]. We should be pleased to receive your consent in accordance with clause 3.7 [*substitute '3.5' when using IC or ICD, '3.3' when using MW, MWD or DB or '62(1)' when using GC/Works/1 (1998)*].

[*List the portions of the Works or design and the names of the sub-contractors.*]

Yours faithfully

Letter 218
To employer, requesting consent to addition of persons to clause 3.8 list
This letter is only suitable for use with SBC.

Dear

In accordance with clause 3.8.2 of the conditions of contract, we should be pleased to receive your consent to the addition of [*insert name*] to the list of persons named in the contract bills [*insert page number and reference*] for [*insert description of work*].

Yours faithfully

Copy: Architect

Letter 219
To employer, giving consent to addition of person to clause 3.8 list
This letter is only suitable for use with SBC.

Dear

Thank you for your letter of the [*insert date*].

We consent to the addition of [*insert name*] to the list of persons named in the contract bills [*insert page number and reference*] for [*insert description of work*].

Yours faithfully

Copy: Architect

Letter 220
To sub-contractor: letter of intent
Special delivery

Dear

Your quotation of the [*insert date*] for [*insert nature of the work or design*] is acceptable and we intend to enter into a sub-contract with you after the main contract documents have been satisfactorily executed.

It is not our intention that this letter, taken alone or in conjunction with your quotation, should form a binding contract. However, we are prepared to instruct you to [*insert the limited nature of the work or design required*]. If, for any reason, the project does not proceed or we instruct you to cease work, our commitment will be strictly limited to payment for the properly executed work you have completed at our request up to the date of our notification that the project will not proceed and/or our instruction to you to cease work. The basis of payment will be the prices in your quotation noted above.

No other work [*substitute 'design' if appropriate*] included in your quotation must be carried out without a further written order. No further obligation is placed upon us under any circumstances.

Yours faithfully

Letter 221
To sub-contractor, regarding part of the construction phase plan
This letter is not suitable for use with GC/Works/SC.

Dear

We draw your attention to those parts of the construction phase plan which are applicable to the sub-contract works and annexed to the schedule of information. The purpose of the plan is to show the way in which the construction phase is to be managed and the important health and safety issues in this project.

Please note that the plan is not to be treated as a mere paper exercise. Rather, it is an important tool in the construction process and something which is a mandatory requirement under the Construction (Design and Management) Regulations 2007.

Copies of the complete plan have been sent to the client and other consultants and relevant parts of the plan to other parties as necessary.

Yours faithfully

Copy: CDM Co-ordinator [*unless the contractor takes this role*]

Letter 222

To sub-contractor, enclosing part of the construction phase plan
This letter is only suitable for use with GC/Works/SC.

Dear

We enclose, for your attention, those parts of the construction phase plan which are applicable to the sub-contract works. The purpose of the plan is to show the way in which the construction phase is to be managed and the important health and safety issues in this project.

Please note that the plan is not to be treated as a mere paper exercise. Rather, it is an important tool in the construction process and something which is a mandatory requirement under the Construction (Design and Management) Regulations 2007.

Copies of the complete plan have been sent to the client and other consultants and relevant parts of the plan to other parties as necessary.

Yours faithfully

Copy: CDM Co-ordinator [*unless the contractor takes this role*]

Letter 223
To domestic sub-contractor, requiring a warranty if not noted in the invitation to tender

Dear

We refer to your quotation dated [*insert date*] in the sum of [*insert amount*] for [*insert the nature of the work or design*].

We are prepared to enter into a sub-contract with you if you will confirm in writing that you are prepared to sign the attached warranty/execute the attached warranty as a deed [*delete as appropriate*] in favour of the employer/future tenants/purchasers/funders [*delete as appropriate*] within two days of receipt of our instruction to do so. This is a requirement of the employer under the main contract and it cannot be varied. If you are not prepared to give the undertaking we seek, we shall have no alternative but to place the sub-contract work elsewhere.

We look forward to receiving your confirmation by [*insert date*] so that we may proceed with the contract documentation.

Yours faithfully

Letter 224
To domestic sub-contractor, requiring a warranty if not noted in the contract documents

Dear

We have received notice from the employer that a warranty is required in favour of [*insert name*], the employer/tenant/purchaser/funder [*delete as appropriate*] and we enclose the relevant warranty.

Please sign/execute as a deed [*delete as appropriate*] and return to us with the contract documents. This is exactly the same form of warranty which was attached to the sub-contract documents.

Please let us have the completed warranty by [*insert date*]. As soon as we receive a copy of the completed warranty from the employer we will send you a copy for your records.

Yours faithfully

Letter 225
To architect, if domestic sub-contractor refuses to provide a warranty which was not originally requested

Dear

[*Insert name of sub-contractor*] refuses to enter into a warranty on the terms which you have proposed.

[*Add either:*]

It appears that this sub-contractor is not prepared to enter into any warranty. Our problem is, as you are aware, that the number of sub-contractors that are able to carry out this kind of work/design [*delete as appropriate*] is extremely limited and this contractor is in great demand. We should be pleased to have your observations and instructions.

[*Or:*]

It appears that this sub-contractor is prepared to enter into a warranty if the terms are amended and we enclose an example of such a warranty. We should be glad to have your confirmation that the amended terms are satisfactory.

Yours faithfully

Letter 226
To architect, objecting to a named person (a)
This letter is only suitable for use with IC or ICD.

Dear

We are in receipt of your instruction number [*insert number*] dated [*insert date*] instructing us to enter into a sub-contract with [*insert name*].

We have reasonable objection, under Schedule 2, paragraph 5.3, to entering into such sub-contract. The reason for our objection is [*explain*].

Yours faithfully

Letter 227
To architect, objecting to a nominated sub-contractor (b)
This letter is only suitable for use with GC/Works/1 (1998).

Dear

We are in receipt of your instruction number [*insert number*] dated [*insert date*] instructing us to enter into a sub-contract with [*insert name*] for [*insert nature of work*] work.

We have reasonable objection to the employment of such nominated sub-contractor, because [*state reasons*]. This objection is made under the provisions of clause 63(6) of the conditions of contract.

Yours faithfully

Letter 228
To architect, if contractor unable to enter into a sub-contract with named person (a)
This letter is only suitable for use with DB.

Dear

In accordance with supplemental provision Schedule 2, paragraph 2.1.1, we have attempted to enter into a sub-contract with [*insert name*] that was named in the Employer's Requirements at [*insert reference to page and item number*]. Our efforts have been unsuccessful, because [*insert reason*] and we should be pleased if you would operate the provisions of paragraph 2.1.2 and either:

1. Issue a change instruction to amend the item in the Employer's Requirements so that we can enter into the sub-contract; or
2. Issue a change instruction to omit the named sub-contract work and issue further instructions about the carrying out of that work.

Yours faithfully

Letter 229

To architect, if unable to enter into sub-contract with named person in accordance with particulars (b)
This letter is only suitable for use with IC or ICD.

Dear

We hereby notify you in accordance with Schedule 2, paragraph 2 of the conditions of contract that we are unable to enter into a sub-contract with [*insert name*] in accordance with the particulars given in the contract documents. The following are the particulars which have prevented the execution of such sub-contract:

[*Specify particulars.*]

We should be pleased if you would issue your instructions as required under the contract.

Yours faithfully

Letter 230
To architect, if some listed sub-contractors will not tender
This letter is only suitable for use with SBC.

Dear

We confirm our telephone conversation earlier today that we have just received the enclosed letters dated [*insert dates*] from sub-contractors listed under clause 3.8, [*insert name*] declining to tender for the work.

Although clause 3.8.3 provides for the addition of further names by agreement or that we should carry out the sub-contract work ourselves, sub-letting with your consent as we see fit, there are few sub-contractors that are capable of doing this kind of work to an acceptable standard.

This is a very serious situation and, if the contract is not to be frustrated or, at the very least, substantially delayed or disrupted, urgent action is required. Please let us have your instructions by return. If you believe a meeting would be useful, please let us know immediately.

Yours faithfully

Letter 231
To architect, if contractor enters into a sub-contract with named person
This letter is only suitable for use with DB.

Dear

This letter is to notify you, as required by supplemental provision, Schedule 2, paragraph 2.1.1, that we entered into a sub-contract with [*insert name*] on [*insert date*].

Yours faithfully

Letter 232
To architect, if contractor enters into contract with named person
This letter is only suitable for use with IC or ICD.

Dear

In accordance with Schedule 2, paragraph 1 of the conditions of contract, we must inform you that we entered into a sub-contract with [*insert name*] on the [*insert date*].

Yours faithfully

Letter 233
To sub-contractor, regarding insurance

Dear

Please submit insurance policies and premium receipts in respect of the insurance which you are required to maintain under clause 6.5 [*substitute '6.5 and 6.10' when using SBCSub/D/C or DBSub/C, '6.5 and 6.14' when using ICSub/D/C or '7' when using GC/Works/SC*] of the sub-contract. The policies and receipts must be in our hands by [*insert date*].

Yours faithfully

Letter 234
To sub-contractor that fails to maintain insurance cover

Dear

We refer to our telephone conversation today with your M.. [*insert name*] and confirm that you are unable to produce documentary evidence that the insurance required by clause 6.5 [*substitute '6.5 and 6.10' when using SBCSub/D/C or DBSub/C, '6.5 and 6.14' when using ICSub/D/C or '7' when using GC/Works/SC*] of the sub-contract has been properly effected and maintained.

In view of the importance of the insurance and without prejudice to your liabilities under clauses 2.5, 6.2 and 6.3 [*substitute '2.2, 2.5, 6.2 and 6.3' when using SBCSub/D/C, '2.4, 6.2 and 6.3' when using ICSub/C, '2.1.1, 2.4, 6.2 and 6.3' when using ICSub/NAM/C or ICSub/D/C or '9.2' when using GC/Works/SC*], we are arranging to exercise our rights under clause 6.5.4 [*substitute '9.2' when using GC/Works/SC*] immediately. Any sum or sums payable by us in respect of premiums will be deducted from any money due or to become due to you or will be recovered from you as a debt.

Yours faithfully

Letter 235
To sub-contractor, enclosing drawings

Dear

In accordance with clause 2.7 [*substitute '2.5' when using ICSub/C, ICSub/NAM/C or ICSub/D/C*] of the sub-contract, [*start here when using GC/Works/SC*] we enclose two copies of each of drawings numbered [*insert numbers*] which we consider to be reasonably necessary to enable you to carry out and complete the sub-contract works.

Yours faithfully

Letter 236
To sub-contractor, if asked to consent to assignment

Dear

Thank you for your letter of the [*insert date*] from which we understand that you wish to assign the sub-contract to [*insert name*].

This is a serious matter and subject to our consent under clause 3.1 [*substitute '26.1' when using GC/Works/SC*]. We cannot make an immediate decision. We are taking advice and we expect to be in a position to write to you again very soon.

Yours faithfully

Letter 237
To sub-contractor that sub-lets without consent

Dear

We are informed that you have sub-let [*insert part of the works sub-let*] to [*insert name*].

Since we have not given our consent, your action is in breach of the sub-contract clause 3.2 [*substitute '26.2' when using GC/Works/SC*] and must cease forthwith. Please confirm, by return, that you will comply with this letter otherwise we will take action to terminate your employment under clause 7.4.1.4 [*substitute '29.1.2' when using GC/Works/SC*].

Yours faithfully

Letter 238
To sub-contractor, giving consent to sub-letting

Dear

In response to your letter of the [*insert date*], we are pleased to give our consent, under the provisions of clause 3.2 [*substitute '26.2' when using GC/Works/SC*] of the sub-contract, to the sub-letting of [*insert portion of the works*] to [*insert name*].

Yours faithfully

Letter 239
To sub-contractor, if no person-in-charge
This letter is not suitable for use with GC/Works/SC.
By fax and post

Dear

At the time of writing there is no person-in-charge of your sub-contract work.
This is in serious breach of your obligation under clause 3.8 [*substitute '3.7'*
when using ICSub/NAM/SC, ICSub/C or ICSub/D/C] to ensure that, at all
times during the execution of such work, you have on site a competent
person-in-charge. Among other things, it is essential that you have
someone on site to receive our directions and to organise your work.

[*If appropriate, add:*]

Currently, you are not proceeding regularly and diligently with your work
which is not being properly executed.

[*Then continue:*]

Clearly, this is having a bad effect on the progress of the main contract Works.
This situation cannot be allowed to continue and if you do not remedy this
major breach by commencement of work on site tomorrow, we shall have no
alternative but to prevent your operatives from working on the site until you
appoint a person-in-charge and notify us of the name. You are aware that any
delay in your work is certain to cause us damages, loss and expense and we
shall recover all such damages from you.

Yours faithfully

Letter 240
To sub-contractor, requiring compliance with direction
Special delivery

Dear

Take this as notice under clause 3.6 [*substitute '4.1' when using GC/Works/SC*] of the conditions of sub-contract that we require you to comply with our direction number [*insert number*] dated [*insert date*], a further copy of which is enclosed.

If within 7 days of receipt of this notice you have not begun to comply, we will employ and pay others to comply with such direction. An appropriate deduction, which in this instance will amount to all costs incurred in connection with such employment, will be taken into account in the calculation of the final sub-contract sum or will be recovered from you as a debt.

Yours faithfully

Copy: Architect

Letter 241
To sub-contractor that fails to comply with direction
Special delivery

Dear

I refer to the notice issued to you on the [insert date] under clause 3.6 [*substitute '4.1' when using GC/Works/SC*] of the conditions of sub-contract requiring compliance with our direction number [*insert number*] dated [*insert date*].

At the time of writing, you have not begun to comply with our direction. We are taking immediate steps to employ and pay others to comply with such direction. All costs incurred in connection with such employment will be taken into account in the calculation of the final sub-contract sum or will be recovered from you as a debt.

Yours faithfully

Copy: Architect

Letter 242
To sub-contractor, if contractor dissents from alleged oral direction
This letter is only suitable for use with SBCSub/C, SBCSub/D/C or DBSub/C.

Dear

We are in receipt of your letter of the [*insert date*] in which you purport to confirm an oral direction which you allege was given by [*insert name*] on site/by telephone [*delete as appropriate*].

We hereby formally dissent from such direction in accordance with clause 3.7 and state that the alleged oral instruction was not given as alleged or at all.

Yours faithfully

Letter 243
To sub-contractor that wrongly confirms an oral direction
This letter is only suitable for use with GC/Works/SC.

Dear

We are in receipt of your letter dated [*insert date*] in which you purport to confirm an oral direction which you allege was given by [*insert name*] on site/by telephone [*delete as appropriate*].

The issue of directions is governed by clause 4.3. We reject your confirmation, because there is no provision for oral directions nor for their confirmation by the sub-contractor.

Moreover, we deny that such a direction was given as you state or at all.

Yours faithfully

Letter 244
To sub-contractor that confirms an oral direction which was given
This letter is only suitable for use with GC/Works/SC.

Dear

We are in receipt of your letter dated [*insert date*] in which you confirm an oral direction which you state was given by [*insert name*] on site/by telephone [*delete as appropriate*].

The issue of directions is governed by clause 4.3 which makes no provision for oral directions nor for their confirmation by the sub-contractor. However, we acknowledge that in this instance such a direction was given and we now attach the direction properly issued in writing.

Please note that we have taken steps to ensure that all future directions will be issued in writing as required under the sub-contract.

Yours faithfully

Letter 245
To sub-contractor, if non-compliant work allowed to remain
This letter is only suitable for use with SBCSub/C, SBCSub/D/C or DBSub/C.

Dear

We write under the provisions of clause 3.12.2 in relation to [*describe the non-compliant work*] which the architect [*substitute 'employer' when using DBSub/C*] has decided not to insist on removal from site.

We confirm that an appropriate deduction which may, but not necessarily will, be the same as any deduction under the main contract, will be taken into account in the calculation of the final sub-contract sum. Alternatively, we may recover it from you as a debt. Please note that, so far as you are concerned, an appropriate deduction will be all the loss and expense we have incurred as a result of the non-compliant work.

Yours faithfully

Letter 246
To sub-contractor, if defective work opened up

Dear

In accordance with clause 3.10 [*substitute '3.9' when using ICSub/NAM/C, ICSub/C or ICSub/D/C or '4.2.1' when using GC/Works/SC*] we instructed you to open up [*describe the work*] for inspection.

The work was found to be not in accordance with the sub-contract and a direction is enclosed under clause 3.11.1 [*substitute '4.3' when using GC/Works/SC*] requiring removal. The cost of opening up and making good is to be entirely at your expense in accordance with clause 3.10 [*substitute '3.9' when using ICSub/NAM/C, ICSub/C or ICSub/D/C or '4.2' when using GC/Works/SC*].

Please take immediate steps to ensure that the work in question is carried out in complete accordance with the sub-contract and that all opening up is made good. Any loss and expense which we have incurred or may incur as a result of your defective work will be deducted from monies which may become due to you.

Yours faithfully

Letter 247
To sub-contractor, inspection after failure of work
This letter is only suitable for use with SBCSub/C, SBCSub/D/C or DBSub/C.

Dear

The [*describe work or materials*] are not in accordance with the sub-contract.

In accordance with clause 3.11.3 of the sub-contract and having due regard to the code of practice in Schedule 1, we enclose our directions for opening up for inspection/testing [*delete as appropriate*] which is reasonable in all the circumstances to establish to our reasonable satisfaction the likelihood or extent of any similar non-compliance.

Whatever the results may be, no adjustment will be taken into account in the calculation of the final sub-contract sum.

Yours faithfully

Letter 248
To sub-contractor, after failure of work
This letter is only suitable for use with ICSub/NAM/C, ICSub/C or ICSub/D/C.

Dear

We have today found that the [*describe the work, materials or goods*] are not in accordance with the sub-contract.

In accordance with clause 3.10.1 we require you to forthwith state in writing what action you propose to immediately take at no cost to us to establish that there is no similar failure in work already executed/materials or goods already supplied [*delete as appropriate*].

Be aware that if you do not respond within 5 days or if we are not satisfied with your proposals or if, because of safety considerations or statutory obligations, we are unable to wait for your proposals, we may issue directions requiring you at no cost to us to open up or test any work, materials or goods to establish whether there is a similar failure including making good thereafter.

Yours faithfully

Letter 249
To sub-contractor, accepting Schedule 2 quotation
This letter is only suitable for use with SBCSub/C, SBCSub/D/C or DBSub/C.

Dear

Thank you for your Schedule 2 quotation dated [*insert date*] which we requested in accordance with clause 5.3 when we issued direction number [*insert number*] dated [*insert date*].

We accept your quotation. Note that you are to carry out the work.

[*Add, as applicable:*]

The amount of the adjustment to the sub-contract sum is [*insert amount*].
The amount to be taken into account in calculating the final sub-contract sum is [*insert amount*].
The amount under paragraph 2.3 is [*insert amount*].
The amount under paragraph 2.4 is [*insert amount*].

[*Add, if applicable, either:*]

The revised period for completion of the sub-contract Works is [insert period].

[*Or:*]

There will be no change to the period for completion of the sub-contract Works.

Yours faithfully

Letter 250
To sub-contractor, rejecting Schedule 2 quotation
This letter is only suitable for use with SBCSub/C, SBCSub/D/C or DBSub/C.

Dear

Thank you for your Schedule 2 quotation dated [*insert date*] which we requested in accordance with clause 5.3 when we issued direction number [*insert number*] dated [*insert date*].

We do not accept your quotation.

[*Add either:*]

The work in our direction is to be carried out and it will be valued under the valuation rules.

[*Or:*]

The work in the direction is not to be carried out.

[*Then add, if the quotation has been prepared on a fair and reasonable basis:*]

A fair and reasonable amount for the cost of preparing the quotation will be included in the final sub-contract sum.

Yours faithfully

Letter 251
To sub-contractor, if necessary to measure work
This letter is only suitable for use with SBCSub/C, SBCSub/D/C or DBSub/C.

Dear

In order to arrive at a valuation, it is necessary to measure [*describe the work*].

The measurement will take place on the [*insert date*] at approximately [*insert time*]. In accordance with clause 5.4 of the sub-contract, we invite you to be present to take such notes and measurements as you may require.

Yours faithfully

Letter 252
To sub-contractor, fixing a new period for completion

Dear

We refer to your notice of delay dated [*insert date and if appropriate add:*] and the further information provided in your letter of the [*insert date*].

In accordance with clause 2.18 [*substitute '2.12' when using ICSub/C, ICSub/NAM/C or ICSub/D/C or '11.5' when using GC/Works/SC*] of the sub-contract, we hereby give you an extension of the period for completion of the sub-contract works of [*insert period*]. The revised period for completion of the sub-contract works is now [*insert period*].

[*When using SBCSub/C, SBCSub/D/C or DBSub/C, add:*]

The relevant sub-contact events taken into account are: [*list events in clause 2.19 and the periods granted in respect of each*].

We have attributed reduction in time to relevant sub-contract omissions as follows: [*list and include the extent of any reduction in extension of time*].

[*When using SC/Works/SC, add:*]

This is an interim/final [*delete a appropriate*] decision.

Yours faithfully

Letter 253
To sub-contractor, fixing a new period for completion after practical completion of the sub-contract works
This letter is not suitable for use with GC/Works/SC.

Dear

In accordance with clause 2.18.5 [*substitute '2.12.3' when using ICSub/C, ICSub/NAM/C or ICSub/D/C*] and after reviewing all the evidence available to us [*add, if appropriate: 'including previous extensions of time'*], we

[*Add:*]

hereby extend the sub-contract period by [*insert additional period*]. The revised period for completion of the sub-contract works is now [*insert period*].

[*Or, when using SBCSub/C, SBCSub/D/C or DBSub/C:*]

hereby shorten the sub-contract period by [*insert the reduction*] having regard to directions for relevant sub-contract omissions. The revised period for completion of the sub-contract works is now [*insert period*].

[*Or:*]

hereby confirm the period for completion previously fixed, namely [*insert period*].

Yours faithfully

Letter 254
To sub-contractor, if no extension of time due

Dear

We have carefully examined your notice of delay [*substitute 'requesting an extension of time' when using GC/Works/SC*] and accompanying particulars dated [*insert date*] and it is our opinion that you are not entitled to an extension of time/further extension of time [*delete as appropriate*] on this occasion.

Yours faithfully

Letter 255
To sub-contractor, if claim for extension of time is not valid

Dear

We refer to your notice of delay [*substitute 'requesting an extension of time' when using GC/Works/SC*] dated [*insert date and if appropriate add:*] and the further information provided in your letter dated [*insert date*].

On the basis of the documents you have presented to us, we see no ground for any extension of time. We shall be pleased to consider any further submissions if they are presented in the proper form and in accordance with the terms of the sub-contract.

Yours faithfully

Letter 256
To sub-contractor, if sub-contract works not complete by due date (a)
This letter is not suitable for use with GC/Works/SC.

Dear

In accordance with clause 2.12 [*substitute '2.15' when using ICSub/C, ICSub/NAM/C or ICSub/D/C*] of the sub-contract, we hereby give notice that the sub-contract [*insert nature of works*] works were not completed within the period for completion/revised period for completion [*delete as appropriate*] ending on [*insert date which should not be unreasonably earlier than the date of this letter*].

You are in breach of the sub-contract and we draw your attention to our right to recover from you the amount of any direct loss and/or expense which we have suffered or incurred as a result of your failure to complete in due time. We reserve all our rights and remedies.

Yours faithfully

Letter 257

To sub-contractor, if sub-contract works not complete within the period for completion (b)
This letter is only suitable for use with GC/Works/SC.

Dear

In accordance with clause 12.1 of the sub-contract, we hereby give notice that the sub-contract [*insert nature of works*] works were not completed within the period for completion/revised period for completion [*delete as appropriate*] ending on [*insert date which should not be unreasonably earlier than the date of this letter*].

You are in breach of the sub-contract and we draw your attention to clause 12.2 and our right to recover from you the amount of any direct loss and/or expense which we have suffered or incurred as a result of your failure to complete in due time. We reserve all our rights and remedies.

Yours faithfully

Letter 258
To sub-contractor, requesting further information in support of a financial claim

Dear

We refer to your claim for loss and/or expense dated [*insert date*].

In order reasonably to enable us to operate the provisions of clause 4.19 [*substitute '4.16' when using ICSub/C, ICSub/NAM/C or ICSub/D/C or '13.1' when using GC/Works/SC*] of the sub-contract, we should be pleased to receive the following information:

[*List the information required.*]

Yours faithfully

Letter 259
To sub-contractor, applying for payment of loss and/or expense (a)
This letter is not suitable for use with GC/Works/SC.

Dear

We hereby give notice and make application under clause 4.21 [*substitute '4.18'
when using ICSub/SC, ICSub/NAM/C or ICSub/D/C*] of the sub-contract as
follows:

We have been caused direct loss and/or expense because the regular progress
of the Works has been materially affected by [*describe*].

Particulars of the calculation of such direct loss and/or expense are enclosed
and we should be pleased to have your agreement by close of business on
[*insert date*] to the amount of [*insert amount*].

Yours faithfully

Letter 260
To sub-contractor, applying for payment of loss and/or expense (b)
This letter is only suitable for use with GC/Works/SC.

Dear

We hereby give notice and make application under clause 12.3 of the sub-contract.

We have been caused direct expense because the regular progress of the Works has been materially affected by the following circumstances [*describe*].

Particulars of the calculation of such direct expense are enclosed and we should be pleased to have your agreement by close of business on [*insert date*] to the amount of [*insert amount*].

In accordance with clause 12.3.3 we confirm that we are entitled to recover expense under clause 12.3.

Yours faithfully

Letter 261
To sub-contractor, giving notice of an interim payment

Dear

This is a written notice specifying the amount of interim payment which is proposed to be made, namely: [*insert amount*]. The amount is calculated on the following basis: [*insert the way in which the amount is calculated. If it is by reference to a certificate, so state*].

Yours faithfully

Letter 262
To sub-contractor, enclosing payment

Dear

In accordance with clause 4.10.4 [*substitute '4.12.3' when using ICSub/C, ICSub/NAM/C or ICSub/D/C or '21.1.1' when using GC/Works/SC*] of the sub-contract, we enclose our cheque in the sum of [*insert amount*]. This is our interim payment number [*insert number*] due on [*insert date*] and calculated as indicated on the enclosed statement.

[*If appropriate, add:*]

Note that we have set-off the sum of [*insert amount*] from money otherwise due to you. Details of the calculation of such sum were sent to you on [*insert date*].

Yours faithfully

Letter 263
To sub-contractor, giving withholding notice.

Dear

Take this as written notice in accordance with clause 4.10.3 [*substitute '4.12.2' when using ICSub/C, ICSub/NAM/C or ICSub/D/C or '21.2.3' when using GC/Works/SC*] that we propose to withhold the sum of [*insert amount*] from the amount notified for payment on [*insert date*].

The grounds for withholding and the amount attributable to each ground are as follows: [*list the grounds with the amount stated for each ground. The list should be as detailed as possible*].

Yours faithfully

Letter 264
To adjudicator, enclosing written statement
Special delivery

Dear

In accordance with your directions, we enclose a written statement setting out particulars of our response to the referral dated [*insert date*] and including the evidence on which we rely.

We are able to provide further information if you so require.

Yours faithfully

Copy: Sub-contractor

Letter 265
To sub-contractor if adjudicator appointed, but there is no dispute
Special delivery and fax

Dear

We note that you have sought the appointment of an adjudicator.

Your notice of intention to refer to adjudication contains no reference to a dispute or difference capable of being referred to adjudication. Therefore, the adjudicator has no jurisdiction. We invite you to withdraw and inform the adjudicator of the position. If you fail to do so, take this as notice that we will reserve all our rights and our participation in the purported adjudication will be without prejudice to our right to resist the enforcement of any decision.

Yours faithfully

Copy: Adjudicator

Letter 266
To adjudicator, if there is no dispute
Special delivery and fax

Dear

The notice of intention to refer to adjudication which was submitted by the referring party on [*insert date*] contains no dispute capable of being referred to adjudication. Therefore, you are lacking jurisdiction and we invite you to relinquish your appointment.

If you fail to do so, take this as notice that we will reserve all our rights and our participation in the purported adjudication will be without prejudice to our right to resist the enforcement of any decision and to seek a declaration from the court that your decision is a nullity and, therefore, you are not entitled to fees.

A copy of the letter dated [*insert date*] which we have sent to the referring party is enclosed.

Yours faithfully

Copy: Sub-contractor

Letter 267
To sub-contractor, if sub-contractor has wrongly sent 7 day notice of intention to suspend performance of obligations

Dear

We are in receipt of your letter of the [*insert date*] which, apparently, you intend to be a notice in accordance with clause 4.11 [*substitute '4.13' when using ICSub/C, ICSub/NAM/C or ICSub/D/C or '24' when using GC/Works/SC*] of the sub-contract.

[*Add either:*]

The purported notice contains a serious error.

[*Or:*]

We are advised that your purported notice is ambiguous.

[*Or:*]

The allegation in your purported notice is incorrect.

[*continued*]

Letter 267 continued

[*Or:*]

The payment to which you refer was made on the [*insert date*].

[*Or:*]

An effective withholding notice was given on the [*insert date*].

[*Then add:*]

Your notice is therefore invalid and of no effect. Take notice that if you suspend further performance of your obligations under the sub-contract, such suspension will be unlawful. We reserve all our legal rights and remedies.

Yours faithfully

Letter 268
To sub-contractor, if sub-contractor has correctly sent 7 day notice of intention to suspend performance of obligations

Dear

We are in receipt of your letter of the [*insert date*] which you have sent as a notice in accordance with clause 4.11 [*substitute '4.13' when using ICSub/C, ICSub/NAM/C or ICSub/D/C or '24' when using GC/Works/SC*] of the sub-contract.

We regret that you have felt it necessary to send such a notice, but we are pleased to enclose our cheque in the sum of [*insert amount*]. We hope you will accept our apologies for this oversight.

Yours faithfully

Letter 269
To sub-contractor, requesting documents for calculation of the final sub-contract sum

Dear

In accordance with clause 4.6.1 [*substitute '21.4.1' when using GC/Works/SC*] of the sub-contract, we should be pleased to receive all documents necessary for the purpose of calculating the final sub-contract sum.

Yours faithfully

Letter 270
To sub-contractor that has failed to submit documents for the calculation of the final sub-contract sum (a)
This letter is only suitable for use with DBSub/C.

Dear

We note that you have failed to submit the documents required under clause 4.6.1 for the purpose of calculating the final sub-contract sum.

It is now more the two months since the practical completion of the sub-contract works and it is my duty under clause 4.6.3 of the sub-contract to prepare a statement of all adjustments to the final sub-contract sum as we can make based on the information we have. A copy of such statement is enclosed.

Yours faithfully

Letter 271
To sub-contractor that has failed to submit documents for the calculation of the final sub-contract sum (b)
This letter is not suitable for use with DBSub/C.

Dear

We note that you have failed to submit the documents required for the purpose of calculating the final sub-contract sum.

The time period allotted in the sub-contract for submission of the documents has now expired and we will now proceed to calculate the final sub-contract sum using the information in our possession.

Your failure to submit the documents is a clear breach of the sub-contract. Although we will calculate the sub-contract sum as fairly as possible, you cannot expect to profit from your breach and, in the absence of appropriate evidence, we shall certainly not make any financial assumptions in your favour.

Yours faithfully

Letter 272
To sub-contractor, enclosing final payment
This letter is not suitable for use with GC/Works/SC.

Dear

We received the final documentation in connection with your final account on [*insert date*]. The final certificate was issued by the architect [*substitute 'final account and final statement became conclusive as to the balance due between the employer and the contractor' when using DBSub/C*] on [*insert date which should be no more than 35 days before the date of this letter*]. We now enclose our cheque in the sum of [*insert amount*]. This sum, which is paid in accordance with clause 4.12.2 [*substitute '4.14.2' when using ICSub/C, ICSub/NAM/C or ICSub/D/C*] of the sub-contract is the final sub-contract sum less only the total amount previously due as interim payments. The enclosed calculations show how we have arrived at such sum.

[*If appropriate, add:*]

Note that we have withheld the sum of [*insert amount*] from money otherwise due to you. A statement of our grounds for so doing and details of the manner in which the amount has been quantified have already been sent to you on [*insert date*].

Yours faithfully

Letter 273

To employer, giving notice of the named sub-contractor's default
This letter is only suitable for use with DB.

Dear

We hereby notify you that, in our opinion, [*insert name*] has made default in [*insert nature of default*] being a matter referred to in clause [*insert clause number as appropriate to the particular sub-contract giving grounds for termination*] of the sub-contract [*insert name of sub-contract form*]. A copy of the clause is attached for your convenience.

Because [*insert name*] is a named sub-contractor under supplemental provision, Schedule 2, paragraph 2, we request your consent under paragraph 2.1.5 to our intention to terminate the sub-contractor's employment.

Yours faithfully

Letter 274
To sub-contractor, giving notice of default before termination
This letter is not suitable for use with GC/Works/SC.
Special delivery

Dear

We hereby give you notice under clause 7.4.1 of the sub-contract that you are in default in the following respect:

[*Insert details of the default with dates if appropriate.*]

If you continue the default for 10 days after receipt of this notice or if you at any time repeat such default, whether previously repeated or not, we may within 10 days of such continuance or within a reasonable time of such repetition terminate your employment under this sub-contract.

Yours faithfully

Letter 275
To sub-contractor, giving notice before determination
This letter is only suitable for use with GC/Works/SC.
Special delivery

Dear

We hereby give you notice under clause 29.1.1 of the sub-contract that the following ground has arisen:

[*Insert details of the ground, which should be contained in clause 29.6.1, 29.6.2 or 29.6.6, with dates if appropriate.*]

If the ground is still in existence for 10 days after receipt of this notice or if it arises again at any time thereafter we may determinate this sub-contract.

Yours faithfully

Letter 276

To architect, if termination of named person's employment possible
This letter is only suitable for use with IC or ICD.

Dear

In accordance with Schedule 2, paragraph 6 of the conditions of contract, we have to advise you that the following events are likely to lead to the termination of [*insert name*]'s employment:

[*Describe the events.*]

We should be pleased to receive your instructions.

Yours faithfully

Letter 277
To sub-contractor, terminating employment after default notice (a)
This letter is not suitable for use with GC/Works/SC.
Special delivery

Dear

We refer to the default notice sent to you on the [*insert date*].

Take this as notice that, in accordance with clause 7.4.2, we hereby terminate your employment under this sub-contract without prejudice to any other rights or remedies which we may possess.

The rights and duties of the parties are governed by clause 7.7. No temporary buildings, plant, tools, equipment, goods or materials shall be removed from site until and if we so direct. Take note that any other provisions which require any further payments or release of retention cease to apply.

Yours faithfully

Letter 278
To sub-contractor, determining the sub-contract after notice (b)
This letter is only suitable for use with GC/Works/SC.
Special delivery

Dear

We refer to the clause 29.1.1 notice sent to you on the [*insert date*].

Take this as notice that, in accordance with clause 29.1, we hereby determine this sub-contract without prejudice to any other rights or remedies which we may possess.

We shall give directions under clause 29.3 in due course.

Yours faithfully

Letter 279
To sub-contractor, determining the sub-contract without prior notice
This letter is only suitable for use with GC/Works/SC.
Special delivery

Dear

The following ground for determination has arisen:

[*Insert details of the grounds which should be contained in clause 29.6 other than clauses 29.6.1, 29.6.2 and 29.6.6.*]

Therefore, take this as notice that, in accordance with clause 29.1, we hereby determine this sub-contract without prejudice to any other rights or remedies which we may possess.

We shall give directions under clause 29.3 in due course.

Yours faithfully

Letter 280
To sub-contractor, terminating employment after termination of the main contract
This letter is not suitable for use with GC/Works/SC.
Special delivery

Dear

In accordance with clause 7.9 of the sub-contract, we have to inform you that our employment under the main contract was terminated on the [*insert date, bearing in mind that this letter must be sent immediately the main contract termination takes place*]. The clause provides that your employment under this sub-contract must thereupon terminate.

The rights and duties of the parties are governed by clause 7.11. Take note that any other provisions which require any further payments or release of retention cease to apply. All our other rights and remedies are preserved.

Yours faithfully

Letter 281
To sub-contractor, termination on insolvency
This letter is not suitable for use with GC/Works/SC.
Special delivery

Dear

We note that you are insolvent as defined in clause 7.1 of the sub-contract. By this letter, we terminate your employment under the sub-contract in accordance with clause 7.5.1.

At our discretion under clause 7.5.3, we may take reasonable measures to ensure that the sub-contract works and the sub-contract site materials are adequately protected.

The rights and duties of the parties are governed by clause 7.7. No temporary buildings, plant, tools, equipment, goods or materials shall be removed from site until and if we so direct. Take note that any other provisions which require any further payments or release of retention cease to apply.

Yours faithfully

Letter 282
To sub-contractor, if cessation of terrorism cover notified
This letter is not suitable for use with GC/Works/SC.

Dear

We have just received notification from [*insert either the name of the insurers or the employer*] that with effect from [*insert the notified cessation date*] terrorism cover will cease and will no longer be available.

We are informing you as required under clause 6.8.1 of the sub-contract and we will let you know as soon as we have details of the employer's election under clause 6.10.2 of the main contract conditions.

Yours faithfully

Letter 283
To sub-contractor, notifying the employer's election after cessation of terrorism cover notified
This letter is not suitable for use with GC/Works/SC.
Special delivery

Dear

Further to our letter dated [*insert date*],

[*Either:*]

we have now received notice that the employer has elected to require the Works to continue to be carried out despite the cessation of terrorism cover. The provisions of clause 6.8.3 of the sub-contract will apply.

[*Or:*]

we have now received notice that the employer has elected to terminate our employment under the main contract on the [*insert date*]. Therefore, in accordance with clause 6.8.2 of the sub-contract, your employment under the sub-contract will also terminate on that date and the provisions of clause 7.11 (except clause 7.11.3.5) will apply.

Yours faithfully

Letter 284
To sub-contractor that serves a default notice (a)
This letter is not suitable for use with GC/Works/SC.
Special delivery

Dear

We are in receipt of your letter dated [*insert date*] which it seems you intended to act as a default notice under clause 7.8.1.

[*Add either:*]

The letter does not comply with the terms of the contract.

[*Or:*]

The letter contains serious errors of fact.

[*Or:*]

The alleged default is the result of your own inadequacies.

[*Or:*]

There is no default such as you allege.

[*Then add:*]

Therefore, your notice is invalid and void of effect. Take this as formal notice that if you persist and attempt to terminate your employment, we shall treat it as a repudiatory breach of contract and take immediate proceedings against you.

Yours faithfully

Letter 285
To sub-contractor that serves a default notice before determination (b)
This letter is only suitable for use with GC/Works/SC.
Special delivery

Dear

We are in receipt of your letter dated [*insert date*] which it seems you intended to act as a notice specifying grounds under clause 30.3.5.

[*Add either:*]

The letter does not comply with the terms of the contract.

[*Or:*]

The letter contains serious errors of fact.

[*Or:*]

The alleged suspension is the result of your own inadequacies.

[*Or:*]

There is no suspension such as you allege.

[*Then add:*]

Therefore, your notice is invalid and void of effect. Take this as formal notice that if you persist and attempt to determine your employment, we shall treat it as a repudiatory breach of contract and take immediate proceedings against you.

Yours faithfully

Letter 286
To sub-contractor that serves notice of termination (a)
This letter is not suitable for use with GC/Works/SC.
Special delivery

Dear

We are in receipt of your letter which purports to be a notice of termination under the provisions of clause 7.8.2 of the sub-contract. [*If appropriate, add:*] We are surprised that you are persisting in what will become a very expensive process for you despite our letter dated [*insert date*] concerning your purported default notice.

[*Add either:*]

The notice is issued prematurely.

[*Or:*]

The notice is not issued in accordance with the terms of the sub-contract.

[*continued*]

Letter 286 continued

[*Or:*]

There is no default.

[*Or:*]

The default was ended on the [*insert date*].

[*Then add:*]

Therefore, we give you 24 hours until close of business on [*insert date*] to withdraw your purported termination and to resume work on site. If you fail to do so, we will take immediate legal advice with a view to accepting your conduct as repudiation of your obligations under the contract. All our other rights and remedies are reserved.

Yours faithfully

Letter 287
To sub-contractor that serves notice of determination (b)
This letter is only suitable for use with GC/Works/SC.
Special delivery

Dear

We are in receipt of your letter which purports to be a notice of determination under the provisions of clause 30.1 of the sub-contract. [*If appropriate, add:*] We are surprised that you are persisting in what will become a very expensive process for you despite our letter dated [*insert date*] concerning your purported notice before determination.

[*Add either:*]

The notice is issued prematurely.

[*Or:*]

The notice is not issued in accordance with the terms of the sub-contract.

[*Or:*]

There is no ground for determination.

[*continued*]

Letter 287 continued

[*Or:*]

The ground for determination was ended on the [*insert date*].

[*Then add:*]

Therefore, we give you 24 hours until close of business on [*insert date*] to withdraw your purported determination and to resume work on site. If you fail to do so, we will take immediate legal advice with a view to accepting your conduct as repudiation of your obligations under the contract. All our other rights and remedies are reserved.

Yours faithfully

Letter 288
To architect, if employment of named person terminated
This letter is only suitable for use with IC or ICD.

Dear

[*If the contractor has advised architect of events likely to lead to termination, begin:*]

Further to our advice of the [*insert date*],

[*Then, or otherwise, begin:*]

We must notify you in accordance with Schedule 2, paragraph 7 of the conditions of contract, that the employment of [*insert name*] was terminated on the [*insert date*]. The circumstances are that [*insert circumstances, quoting clause numbers and dates as appropriate*].

We should be pleased to receive your instructions in accordance with paragraph 7.

Yours faithfully

Letter 289
To architect, if employment of named person terminated under ICSub/NAM/C clause 7.4, 7.5 or 7.6
This letter is only suitable for use with IC or ICD.

Dear

We refer to the termination of the [*insert nature of work*] sub-contract employment of [*insert name*] on [*insert date*] under the provisions of ICSub/NAM/C clause 7.4/7.5/7.6 [*delete as appropriate*]. In accordance with the provisions of Schedule 2, paragraph 10.2 of the conditions of main contract, we have taken reasonable action to recover additional amounts payable to us by the employer as a result of the application of paragraph 8.1/8.2/9 [*delete as appropriate and add, if appropriate:*] together with an amount equal to the liquidated damages that would have been payable or allowable by us to the employer under clause 2.23 of the main contract, but for the application of Schedule 2, paragraph 8.1/8.2/9 [*delete as appropriate*]. The total of such amounts is [*insert total*] made up as follows: [*insert a breakdown of the amount*].

[*Add either:*]

We attach copies of our letters of the [*insert dates*] to [*insert name*] which we consider to be reasonable action.

[*continued*]

Letter 289 continued

[*Or:*]

We attach copies of our letters dated [*insert dates*] to [*insert name*] which we consider to be reasonable action. Also enclosed is a copy of a letter from [*insert name*] dated [*insert date*] from which you will see that they dispute their liability to pay. Although we are prepared to continue our efforts to secure payment, we do not consider that our efforts are likely to meet with success. Please inform us if you require us to commence adjudication, arbitration or other proceedings in respect of [*insert name*]. We are prepared to commence such proceedings only if the employer agrees in writing to indemnify us against any legal costs in accordance with the provisions of Schedule 2, paragraph 10.2.3.

Yours faithfully

Copy: Employer

Letter 290
To architect, if contractor instructed to carry out named person's work after termination of named person's employment
This letter is only suitable for use with IC or ICD.

Dear

We are in receipt of your instructions under the provisions of Schedule 2, paragraph 7.2 of the conditions of contract.

This clause permits us to sub-contract the [*insert nature of work*] work if we so wish. We are giving the matter our urgent consideration and if we decide to so sub-let, we will write to you again.

Yours faithfully

[*Note that a paragraph 7.2 instruction ranks as a clause 2.19 relevant event, a clause 4.17 relevant matter and as a variation.*]

Letter 291
To architect, if contractor decides to sub-let after termination of named person's employment
This letter is only suitable for use with IC or ICD.

Dear

Further to our letter of the [*insert date*], we have decided to sub-let the [*insert nature of work*] work to [*insert name and address of the sub-contractor*].

[*If appropriate, add:*]

You may be aware that we do not usually carry out work of this type ourselves and sub-letting to [*insert name of sub-contractor*] represents the best way of dealing with the matter. However, there may be some difficulties with regard to programming which we are trying to resolve.

Yours faithfully

[*Note that a paragraph 7.2 instruction ranks as a clause 2.19 relevant event, a clause 4.17 relevant matter and as a variation.*]

Letter 292
To employer, if money recovered after termination of named person's employment
This letter is only suitable for use with IC, ICD or DB.

Dear

We refer to our letter of the [*insert date*] regarding our efforts to recover amounts from [*insert name*] in accordance with Schedule 2, paragraph 10.2.1 [*substitute 'supplemental provision, Schedule 2, paragraph 2.1.7' when using DB*].

We are pleased to report that our efforts on your behalf have been reasonably successful and we have recovered the sum of [*insert amount*] which is made up as follows: [*insert breakdown of amount*].

Yours faithfully

Letter 293
To sub-contractor, requesting details for the health and safety file

Dear

You have an obligation to provide details for inclusion in the health and safety file. We require such details <u>so that we can pass them to the CDM Co-ordinator</u> before practical completion [*omit 'practical' when using GC/Works/1 (1998) and delete the underlined phrase when using DB*].

You should note that practical completion [*omit 'practical' when using GC/Works/1 (1998)*] and all that implies will not be achieved until you have complied with this obligation.

Yours faithfully

Letter 294
**To sub-contractor, if contractor dissents from date of
practical completion**

Dear

We are in receipt of your notification dated [*insert date which should not be more
the 14 days before the date of this letter*] stating that, in your opinion, practical
completion [*substitute 'completion' when using GC/Works/SC*] of the sub-contract
[*insert nature of works*] works will be achieved on [*insert date*].

Take this as formal notice, in accordance with clause 2.20.1 [*substitute
'2.14.1' when using ICSub/C, ICSub/NAM/C or ICSub/D/C or '14.1' when using
GC/Works/SC*] of the sub-contract, that we dissent from such date of practical
completion [*substitute 'completion' when using GC/Works/SC*] for the following
reasons:

[*State reasons clearly, but briefly.*]

Yours faithfully

Letter 295
To sub-contractor, notifying practical completion after earlier dissent

Dear

Take this as notice under clause 2.20.2 [*substitute '2.14.2' when using ICSub/C, ICSub/NAM/C or ICSub/D/C or '14.2' when using GC/Works/SC*] of the sub-contract that practical completion [*substitute 'completion' when using GC/Works/SC*] of the sub-contract works took place on the [*insert date*]. Practical completion [*substitute 'completion' when using GC/Works/SC*] shall be deemed for all the purposes of the sub-contract to have taken place on that date.

Yours faithfully

Letter 296
To sub-contractor, enclosing schedule of defects

Dear

The rectification [*substitute 'maintenance' when using GC/Works/SC*] period in the main contract ended on the [*insert date*] and we have received a schedule of defects from the architect.

Certain of the defects are your responsibility under clause 2.22 [*substitute '2.16' when using ICSub/C, ICSub/NAM/C or ICSub/D/C or '14.3' when using GC/Works/SC*] of the sub-contract and we enclose a schedule of such defects for your attention. Please carry out the necessary remedial work forthwith.

Yours faithfully

Letter 297
To sub-contractor, directing that some defects are not to be made good

Dear

The rectification period in the main contract ended on the [*insert date*] and we have received a schedule of defects from the architect.

Certain of the defects are your liability under clause 2.22 [*substitute '2.16' when using ICSub/C, ICSub/NAM/C or ICSub/D/C or '14.3' when using GC/Works/SC*] of the sub-contract and we enclose a schedule of such defects for your attention.

[*Either:*]

The architect has instructed that you are not required to make good any of the defects shown on the schedule.

[*Or:*]

The architect has instructed that you are not required to make good those defects marked 'E'.

[*Then:*]

We have been informed that an appropriate deduction will be made from the main contract sum in respect of defects which are not required to be made good. You must bear an appropriate proportion of such deduction and it will be taken into account in the calculation of the final sub-contract sum or it will be recoverable from you as a debt.

Yours faithfully

Letter 298
To sub-contractor, if deduction made under main contract for inaccurate setting out
This letter is not suitable for use with GC/Works/SC.

Dear

We have been notified under the main contract, clause 2.10 [*substitute '2.9' when using IC or ICD or '2.35' when using DB*], that the employer is deducting the sum of [*state the sum in words and figures*] from the contract sum due to inaccurate setting out.

The setting out in question is your responsibility and, in accordance with clause 2.23 [*substitute '2.17' when using ICSub/NAM/C, ICSub/C or ICSub/D/C*] of the sub-contract, it will be taken into account in the calculation of the final sub-contract sum or it will be recoverable from you as a debt.

Yours faithfully

Letter 299

To architect, if action threatened because of named person's design failure
This letter is only suitable for use with IC or ICD.

Dear

Thank you for your letter of the [*insert date*].

We note that the employer is seeking to hold us responsible for a design defect in the sub-contract [*insert nature of work*] works designed and carried out by [*insert name*].

Our liability in respect of such defect is expressly excluded by Schedule 2, paragraph 11 of the contract.

Yours faithfully

Copy: Employer

Letter 300
To sub-contract architect, engineer or other consultant, regarding professional indemnity insurance
This letter is only suitable for use with SBC, ICD, MWD or DB.

Dear

We refer to our telephone conversation of [*insert date*] regarding the possibility that we shall employ you to carry out certain design functions in relation to this project. We confirm that we are engaged under the [*insert the name of the contract as shown on the cover*] by the employer, [*insert name*].

We are considering the fee proposal you put to us on the [*insert date*] but before we can take the matter further, we should be pleased to receive evidence that you maintain adequate and suitable professional indemnity insurance and that you will continue to maintain such insurance for a period of [*insert period*] after practical completion of the Works.

Yours faithfully

Letter 301
To sub-contract architect, engineer or other consultant, if late in providing information
This letter is only suitable for use with SBC, ICD, MWD or DB.

Dear

This is to inform you that we are not receiving the information, which it is your responsibility to prepare, within the appropriate time periods to enable us to place our orders/make the components/carry out the work [*delete as appropriate*]. The following schedule speaks for itself:

[*List the descriptions of drawings requested with dates and the dates actually supplied.*]

You will recall that we jointly prepared the drawing issue schedule on [*insert date*] before work began on site. Your delays are causing us expense and possibly we may be liable for liquidated damages under the main contract. We certainly have no grounds for recovery of our loss from the employer. To be frank, your delay in providing us with information in conformity with the jointly agreed schedule is a breach of contract for which we are entitled to recover damages. For the moment we are reserving all our rights and remedies in that regard.

Take note that it is essential that we receive all the delayed information by [*insert date*] at latest and then continue to receive further information on the due dates. If there are any subsequent delays in receipt of information, we shall be obliged to set-off our loss and expense against any money due or to become due to you.

Yours faithfully

Letter 302
**To sub-contract architect, engineer or other consultant, if action
threatened because of design failure**
This letter is only suitable for use with SBC, ICD, MWD or DB.

Dear

We have received a letter from the employer dated [*insert date*], a copy of
which is enclosed. You will note that the employer is seeking to hold us
responsible for a defect in the Works. The defect in question appears to be
caused by a failure of design.

This part of the work is your responsibility and we have arranged a meeting
on site/at our offices [*delete as appropriate*] on [*insert date*] at [*insert time*] to
examine the problem and achieve a solution and we should be pleased to have
your confirmation that you will attend.

Yours faithfully

Letter 303
To sub-contract architect, engineer or other consultant, at the end of a successful project
This letter is only suitable for use with SBC, ICD, MWD or DB.

Dear

We enclose our final payment in the sum of [*insert amount*] in respect of this project.

Despite the usual problems on this scale of work, all is now complete and the building looks well and, apparently, functions to the employer's satisfaction. Indeed the employer seems extremely pleased with the outcome.

With the employer's permission, we are arranging to photograph the building for use in our publicity material where, of course, we shall give you proper acknowledgement. We presume, unless we hear to the contrary, that you have no objection to this course of action. We shall, of course, send you copies of the photographs and of the publicity material.

We have enjoyed working with you on this project and we look forward to co-operating again on future projects.

Yours faithfully

Index

Printed and bound by CPI Group (UK) Ltd, Croydon, CR0 4YY

27/10/2024

14580199-0003